Electrical Manipulation of Cells

Electrical Manipulation of Cells

edited by

Paul T. Lynch
University of Derby

Michael R. Davey
University of Nottingham

CHAPMAN & HALL

 New York • Albany • Bonn • Boston • Cincinnati • Detroit • London • Madrid • Melbourne
Mexico City • Pacific Grove • Paris • San Francisco • Singapore • Tokyo • Toronto • Washington

Art direction: Andrea Meyer, emDASH inc.
Cover design: Saeed Sayrafiezadeh, emDASH inc.

Copyright © 1996
Chapman & Hall

Printed in the United States of America

For more information, contact:

Chapman & Hall
115 Fifth Avenue
New York, NY 10003

Thomas Nelson Australia
102 Dodds Street
South Melbourne, 3205
Victoria, Australia

Nelson Canada
1120 Birchmount Road
Scarborough, Ontario
Canada, M1K 5G4

International Thomson Editores
Campos Eliseos 385, Piso 7
Col. Polanco
11560 Mexico D.F. Mexico

Chapman & Hall
2-6 Boundary Row
London SE1 8HN
England

Chapman & Hall GmbH
Postfach 100 263
D-69442 Weinheim
Germany

International Thomson Publishing Asia
221 Henderson Road #05-10
Henderson Building
Singapore 0315

International Thomson Publishing-Japan
Hirakawacho-cho Kyowa Building, 3F
1-2-1 Hirakawacho-cho
Chiyoda-ku, 102 Tokyo
Japan

1 2 3 4 5 6 7 8 9 10 XXX 01 00 99 97 96

Library of Congress Cataloging-in-Publication Data

Electrical manipulation of cells / edited by P.T. Lynch and M.R. Davey.
        p.   cm.
    Includes bibliographical references and index.
    ISBN 0–412–03001–2 (alk. paper)
    1. Electroporation.   2. Bioelectrophysiology.   I. Lynch, P.T. (Paul Thomas).
II. Davey, M.R. (Michael Raymond).
    [DNLM: 1. Electroporation—methods.   2. Cell Fusion—physiology.   3. Electric
Stimulation—methods.   QH 585 E39 1995]
QH586.5.E48E43—1996
574.87′041—dc20   96–10508
                                                                              CIP

British Library Cataloguing in Publication Data available

To order for this or any other Chapman & Hall book, please contact **International Thomson Publishing,
7625 Empire Drive, Florence, KY 41042.** Phone: (606) 525-6600.
Fax: (606) 525-7778, e-mail: order@chaphall.com.

For a complete listing of Chapman & Hall's titles, send your requests to
Chapman & Hall, Dept. BC, 115 Fifth Avenue, New York, NY 10003.

Contents

Preface

The influence of electrical fields on living matter represents an important meeting area for the physical and biological sciences, where there remains on-going fundamental research. This volume includes a range of topics from examination of the mechanisms of action of external electrical fields on biological material, to the ways in which electrical stimuli are employed to manipulate cells. The relevance of such electrical approaches is discussed in relation to present-day biotechnology.

Traditionally, in biophysics, electric fields have been used to study the ionic-electrical properties of molecules and molecular groupings, such as biological membranes. External electric fields exert so-called "ponderomotoric" forces on biological particles and cells. These forces may result in cell rotation, which can be employed to probe cellular properties such as membrane capacitance, cytoplasmic conductivity and cell movement. Electrically stimulated cell movement will separate different cell types or bring them together, via dielectrophoresis, to form "pearl chains", the latter being the first stage in cell fusion.

The ability of externally applied electrical fields to induce the formation of pores in biological membranes is a key feature in the electrical manipulation of cells. Electroporation-induced transient loss of the fundamental barrier function of biological membranes provides a means by which large molecules such as DNA, which are normally excluded by the plasma membrane, may enter cells. The electroporation of DNA into cells, integration of DNA into the genome and its subsequent expression in recipient cells, provides an experimental approach for genetically modifying organisms. In addition, electroporation is also a potential means of drug delivery in chemotherapy. A combination of dielectrophoresis and electroporation forms the basis of electrofusion methodologies, for combining the cytoplasmic and genomic components of cells. In this respect, electrofusion has been employed to generate, for example, animal hybridoma cells and plant somatic hybrids and cybrids.

Exposure of cells to weak, long-term electric fields, to high voltage short duration pulses and to magnetic fields, influences cell development. Some of these electrical treatments may have medical applications, as in the case of the reported stimulation of bone regrowth. In plants, electrical parameters also exert pronounced effects on cell division and, more importantly, shoot regeneration from cultured tissues. The mechanisms involved in the stimulation of growth of animal and plant cells are not fully understood. However, the effects in plants may be manifest, in part, through alteration in the endogenous growth regulators in electrically treated tissues.

The chapters in this book encompass a range of topics, each of which is presented as a review. Overall, the volume will be of interest to students and research workers in the animal, medical and plant sciences.

P. T. Lynch
University of Derby
M. R. Davey
University of Nottingham

Contributors

F. J. Ascencor
Departamento de Física Aplicada
Universidad del País Vasco
Bilbao, Spain

S. B. Dev
Genetronics Inc.
11199-A Sorrento Valley Road
San Diego, CA 92121-1334
USA

N. V. Blackhall
Plant Genetic Manipulation Group
Department of Life Science
University of Nottingham
University Park
Nottingham NG7 2RD
UK

A. Domínguez
Departamento de Microbiología
y Genética
Universidad de Salamanca
Salamanca, Spain

J. Brunstedt
Danisco Biotechnology
Højbygårdvej 14
DK-4960 Holeby
Denmark

G. Fuhr
Institute for Biology
Department of Membrane Physiology
Humboldt-University of Berlin
Invalidenstr. 42
D-10115 Berlin, Germany

M. R. Davey
Plant Genetic Manipulation Group
Department of Life Science
University of Nottingham
University Park
Nottingham NG7 2RD
UK

A Goldsworthy
Department of Biology
Imperial College of Science,
    Technology and Medicine
London SW7 2BB
UK

R. Hagedorn
Institute for Biology
Department of Membrane Physiology
Humboldt-University of Berlin
Invalidenstr. 42
D-10115 Berlin, Germany

R. Heller
Department of Surgery
College of Medicine
MDC Box 16
University of South Florida
12901 Bruce B. Downs Boulevard
Tampa, FL 33612-4799
USA

G. A. Hofmann
Genetronics Inc.
11199-A Sorrento Valley Road
San Diego, CA 92121-1334
USA

F. J. Iglesias
Departamento de Física Aplicada
Universidad de Salamanca
Salamanca, Spain

M. Jaroszeski
Department of Chemical Engineering
College of Engineering
University of South Florida
Tampa, FL 33620
USA

M. Joersbo
Danisco Biotechnology
Højbygårdvej 14
DK-4960 Holeby
Denmark

B. Jones
Plant Genetic Manipulation Group
Department of Life Science
University of Nottingham
University Park
Nottingham NG7 2RD
UK

D. B. Jones
Experimental Orthopaedics
Philipps-Universität
Baldingerstr.
35033 Marburg
Germany

M. G. K. Jones
Plant Biotechnology Laboratory
School of Biological
    and Environmental Sciences
Murdoch University
Perth, Western Australia 6150

K. C. Lowe
Plant Genetic Manipulation Group
Department of Life Science
University of Nottingham
University Park
Nottingham NG7 2RD
UK

J. A. Lucy
The Babraham Institute
Babraham Hall
Babraham
Cambridge CB2 4AT
UK

P. T. Lynch
Plant Biotechnology Group
Division of Biological Sciences
University of Derby
Kedleston Road
Derby DE22 1GB
UK

J. B. Power
Plant Genetic Manipulation Group
Department of Life Science
University of Nottingham
University Park
Nottingham NG7 2RD
UK

B. McLeod
Department of Electrical Engineering
Montana State University
Bozeman, Montana
USA

C. Santamaría
Departamento de Física Aplicada
Universidad del País Vasco
Bilbao, Spain

E. Neumann
Faculty of Chemistry
University of Bielefeld
P. O. Box 100131
D-33501 Bielefeld
Germany

T. Y. Tsong
Department of Biochemistry
University of Minnesota College
   of Biological Sciences
St. Paul, Minnesota 55108
USA

CHAPTER 1

Electrofusion and Electroporation Equipment

B. Jones

P. T. Lynch

J. B. Power

M. R. Davey

ABSTRACT

This chapter discusses ways of manipulating electrical output to permeabilize and to fuse cells and provides a historical summary of the development of electromanipulation equipment. It also discusses the merits and limitations of both commercially and noncommercial systems.

INTRODUCTION

The use of electrical methods to manipulate living cells genetically by electroporation and electrofusion has resulted in the production of a range of commercially available pulse generators, together with the development of noncommercial equipment. An important prerequisite in the choice of any instrumentation is a basic understanding of the electrical mechanisms that underlie the functioning of such equipment and of the possible effects of electrical pulses on the biological system under investigation.

1

Electroporation

The formation of pores in the plasma membrane is induced by the application of electric fields, normally in the form of a direct current (DC) applied across an electrode chamber in which the living cells are suspended. Due to electrostatic induction, ion migration occurs within cells located between, but not in direct contact with, two oppositely energized electrical conductors (electrodes). These charges within the cells move along the electrical lines of force to take up energetically favorable positions within the cells; that is, they separate and form charged poles within the cells along the electrical field lines, corresponding to oppositely charged electrode plates.

Three electrical parameters are used to control electroporation; namely, the electrical field strength, the pulse duration, and the number of pulses applied. Cell lysis may ensue if one of these parameters is too high or if the osmotic potential of the electroporation solution is incorrect. The electrical field strength (E) is the force with which the field acts on a unit charge situated at a particular point in space and is determined by the applied electrical potential (voltage, V) and the distance (d) between the electrodes; that is, $E = V/d$. If the distance is measured in centimeters (cm), then the electrical field strength is given as V/cm.

The number and duration, ranging from tens of microseconds (μs) to a few milliseconds (msec), of the applied electrical pulses can affect the number and size of the resultant pores in biological membranes. A certain critical field strength, E_{cr}, however, must be reached before pore formation will occur. Theoretically, the field strength required to produce a localized breakdown of the membrane at the field-induced poles can be calculated from the equation $E_{cr} = V_{cr}/1.5r$, where V_{cr} is the transmembrane potential difference required to induce localized breakdown of the membrane and r is the radius of the cell. V_{cr} values of 0.5 to 1.5 V have been found, according to the cell type and the experimental conditions (Bates et al., 1987). Depending on the state of the membrane and the external environment such as the temperature, electroinduced membrane pores may remain open for seconds or hours (Kinosita and Tsong, 1977a, 1977b; Zimmermann and Vienken, 1982; Glaser et al., 1986) and exogenously located macromolecules such as DNA may cross the cell membrane (electrotransfection). The molecules to be transferred are usually included in the electroporation solution before application of the electric field.

The term electroporation, although commonly used when referring to electrotransfection, is also a fundamental event in the electrofusion process. When electroporation occurs in two or more closely aligned cells, the membranes of the adjacent cells may collapse at their points of contact, followed by cytoplasmic mixing to instigate cell fusion (Zimmermann and Vienken, 1982).

Electrofusion

Cell Alignment

Electrofusion is achieved using an electroporator, together with one of several methods of bringing cells into close membrane–membrane contact. Such methods include the natural adhesion of slime moulds (Neumann et al., 1980), the physical juxtapositioning of cells by manipulation with electrodes (Senda et al., 1979; Zimmermann and Scheurich, 1981), chemical agglutination (Weber et al., 1981; Chapel et al., 1986), sedimentation at high cell density to produce tight monolayers (Sowers, 1985) or multiple layers (Teissié and Rols, 1986), the formation of monolayers through cell growth (Teissié et al., 1982), immunological targeting (Lo et al., 1984; Lo and Tsong, 1989), and cell polarization induced by AC or magnetic fields with resulting mutual dielectrophoresis (Pohl, 1978; Schwan, 1989; Zimmermann et al., 1985).

Mutual Dielectrophoresis

Pohl (1951, 1958) presented a detailed description of the movement of neutral particles under the influence of nonuniform AC fields. In an electrical field, cells become polarized and the resulting cellular dipoles are attracted toward the oppositely charged electrodes. In a homogeneous field, the attractive forces acting on each side of a cell are balanced. A nonhomogeneous electric field may however, induce translational movement of the cell directed toward the pole at which the net attractive force is strongest. The electrical field will be homogeneous if the electrodes are perfectly flat, equidistant from one another and continuous across the opposite walls of the fusion chamber. When cells are introduced into the fusion chamber, however, localized nonhomogeneity in the field caused by the induced cellular dipoles will, if of sufficient strength, induce the translational movement of cells toward each other, the positive pole of one cell being attracted toward the negative pole of another. This phenomenon, termed mutual dielectrophoresis (Pohl, 1978; Schwan, 1989), leads to the formation of rows of cells, often referred to as pearl chains, along the lines of force. When cell-to-cell contact has been achieved, single or multiple high-intensity electrical pulses will induce membrane breakdown. If applied simultaneously, the AC and DC wave-forms will conflict and partially cancel each other. Therefore, the integration of the alignment (AC) field and fusion (DC) field is normally achieved by momentarily switching between the two outputs.

Commercial and Noncommercial Equipment

Several commercially produced instruments are available but frequently at a prohibitive cost. Furthermore, some biological systems require a degree of control and flexibility that cannot be obtained from standard, commercially manufactured units. Several authors have described electrofusion equipment of simple design that is inexpensive to construct (Watts and King, 1984; Zachrisson and Bornman, 1984; Mischke et al., 1986; Kramer et al., 1987). The facilities and parameters provided by such equipment are however, generally limited. In contrast, Jones et al. (1994) have reported the construction of an instrument that provides a considerable degree of flexibility and control over the fusion process. This apparatus has proven useful in the large-scale electrofusion of isolated higher plant protoplasts in somatic hybridization and in the electrotransfection of plant protoplasts resulting in the production of transgenic plants.

EQUIPMENT DESIGN

Electroporation

Many of the devices used for electroporation use the full discharge of one or more capacitors to provide a DC pulse (Fromm et al., 1985; Rech et al., 1989; Sowers, 1984, 1989). The advantages of full capacitive discharge (FCD) devices are their simple construction, low cost and ability to deliver high voltages. Rech et al. (1989) described the components required and the circuitry to construct a simple electroporator. When connected to a power supply, which in the case of noncommercial equipment may be an electrophoresis power supply, a capacitor builds up an electrical charge. The latter is limited by the energy storage potential of the capacitor. The current (voltage) released by a capacitor is related to the speed at which the stored electrical energy is allowed to dissipate and not to the voltage used to charge the capacitor. Thus, if the discharging energy meets a high resistance, it is released slowly, resulting in a relatively long pulse duration at low voltage. To allow the electrical pulse to terminate naturally when the stored energy is released, the capacitor is isolated from its charging source before being discharged. The electrical resistance across the electrode assembly at the time the pulse is applied affects the rate at which the capacitor is discharged and, consequently, influences the pulse profile. Even if the ionic content of the solution bathing the cells is precisely controlled, inconsistencies in resistance can arise because of the release of ions from the cells with time or of cellular damage that may result during sample preparation. This prevents the exact duplication of conditions. Consequently, the applied pulse parameters may vary from experiment to experiment.

The pulse provided by a discharging capacitor is usually exponentially decaying, although a square pulse can be attained with some electronic enhancement (Saunders et al., 1989). Unlike capacitive discharge devices, the DC pulse length provided by a logic driven unit (microprocessor controlled) is independent of the resistance of the electroporation cell, thus ensuring electrical reproducibility regardless of any variation in conductivity of the solution bathing the cells. The value of rectangular DC wave forms, compared to the sinusoidal pulse profile normally obtained from a FCD device, has been debated, especially in the context of protoplast viability (Saunders et al., 1989; Joersbo and Brunstedt, 1990). Jones et al. (1993) reported low cell viability following gene uptake into isolated protoplasts of the leguminous plants *Glycine argyrea*, *Medicago sativa*, *M. varia*, and *Stylosanthes macrocephala* when the protoplasts were electroporated in various ionic solutions with a noncommercial FCD instrument built according to the circuit diagrams given by Rech et al. (1989). In contrast, using a noncommercial logic-controlled electroporator to generate electronically defined square DC pulses, Jones et al. (1993) found that viability was conserved in protoplasts electroporated in a solution of low ionic strength consisting of mannitol, the osmoregulator used to prevent protoplast lysis and 0.5 mM calcium chloride for membrane stabilization. Additionally, the uptake of plasmid DW2 (Pietrzak et al., 1986) into isolated protoplasts of *Glycine*, *Medicago* and *Stylosanthes* using the logic-controlled electroporator (Jones et al., 1994) resulted in strong transient expression of the chloramphenicol acetyltransferase (*cat*) reporter gene in protoplasts of these three genera.

Although a different wave-form was applied in the above investigations by Jones et al. (1993), the composition of the electroporation solution may have contributed, to some extent, to the differences in protoplast viability. When influenced by an electrical field, ion migration occurs within the cells, causing charge polarization. It is the electrical field that induces cell polarization, not the flow of electrical current through the solution in which the cells are suspended. The conductivity of the cell membrane, which in its natural state is an effective electrical insulator, increases during electropermeabilization. Therefore, if electroporation solutions with substantial ionic contents are used, as is often the case, the cytoplasm of the cell may be exposed to potentially lethal electrical currents.

Molecule oscillation, produced by high-intensity, pulsed AC fields, can also induce membrane disruption (Chang, 1989). Joersbo and Brunstedt (1990) designed a simple AC electroporator, consisting of a plug, a length of 13-A electrical cable, a fuse holder, and an electrode. The electrode assembly, containing a suspension of protoplasts, was connected to the main electrical supply of the laboratory by a fuse. Switching on the mains caused the fuse to blow, an electrical pulse being delivered to the electrode during the time taken for the fuse to blow. Since the rating of the fuse determined the time taken for the fuse to blow, it was possible to vary the pulse duration by using fuses of different values. This simple arrangement was used to induce gene intake into protoplasts of sugarbeet, but it has severe limitations in relation to operator safety.

Electrodes

Electrodes supplied with commercially manufactured electroporation equipment normally consist of a sealable cylinder with an electrode plate at each end, as in the Dia-log apparatus (Dia-log, Harffstr. 3, Düsseldorf, Germany). Circular or parallel plate electrodes of stainless steel are also available (e.g., Hoefer Scientific Instruments) and are designed to fit the wells of commercially available plastic multiwell dishes and spectrophotometer cuvettes. An electrode commonly used in conjunction with non-commercial electroporation equipment consists of a 1 mL capacity plastic cuvette, lined on two opposite internal walls with aluminum foil, the latter being secured in position by double-sided tape (Rech et al., 1989). Continuing each strip of foil down the corresponding outer wall of the cuvette permits contact with the pulse generator through sprung electrodes in the cuvette support stand. An important addition is a microswitch set into an enclosing cover to protect the user from accidental electric shock.

Electrofusion

The electrical resistance of the solution bathing the cells is an important consideration with electrical pulse generators, as the internal resistance of the device should match the resistance of the output load, the latter being the chamber containing the cell suspension. Additionally, heating can occur when the conductivity of an electrofusion solution is high, as is the case when the solution has a high ionic content. Heating may occur during electroporation, but because of the short period of exposure, it is less significant than with electrofusion, where additional, sustained electrical currents are used to induce cell alignment. Therefore, electrofusion is normally performed in a solution of low ionic strength, with only small quantities of calcium chloride or magnesium sulfate to aid membrane stability. Blangero and Teissié (1985) have discussed, at length, the effects of the composition of electrofusion solutions on the yield of fusion products.

Fusion Pulse (DC)

Many of the noncommercial electrofusion units in use are FCD based. A simple FCD system has trigger-activated mercury-wetted electrodes (Sowers, 1984) or Thyratons (Sowers, 1989) that synchronously switch between an AC power supply and a charged capacitor. Sowers (1989) designed a partial capacitive discharge circuit that generated pulses by using the AC cycle (60 Hz) to open and close a pair of mercury-wetted electrodes at 60 times per second. It was possible to generate a timed (partial) capacitive discharge with a rectangular pulse profile by introducing a phase shift to alter the duration of each on–off sequence (pulse) and a latching relay to allow only one pulse to connect to the chamber. It is important that mercury-wetted relays or Thyratons are used to switch between AC and DC outputs, since contacts in conventional relays tend to bounce, resulting in erratic and inconsistent DC pulses.

An alternative and less complex way of constructing an electrofusion system capable of providing electrically clean and consistent wave-forms of both direct and indirect currents is by linking together commercially available sine wave generators. Mischke et al. (1986) used a junction box and mercury-wetted relay to control and to coordinate the output of a function generator (AC) and a pulse generator (DC). Given some knowledge of electronics, it is possible to achieve a similar but less expensive facility by using integrated circuits. Due to low-operating voltages, however, the output obtained directly from integrated circuits provides little power at the electrode. An integrated circuit-based device, designed to facilitate practical demonstrations of electromanipulation of plant protoplasts, was reported by Kramer et al. (1987). Alignment was achieved using an Astable Multivibrator integrated circuit at frequencies of 100 kHz to 3 MHz, with a maximum voltage of 7.5 V. The duration of an applied DC fusion pulse, provided by an electrophoresis power supply (50 V maximum), was controlled by a second integrated circuit (Monostable Multivibrator). The electrical field strengths required to obtain fusion were obtained using a narrow gap between the electrodes. Although this inexpensive equipment is unsuitable and is not intended for large-scale electromanipulation, it is ideal for simple observations of the influence of electrical fields on cells and, because of the low-voltage output, is safe for student use.

An optical bridge arrangement (optoisolator) enabled Jones et al. (1994) to use integrated circuit technology to define multiple square wave DC pulses and to create enough power to drive an electrical field across widely spaced electrodes. Detailed plans together with a list of components for the construction of this versatile instrument are given by the authors. The output of the low-voltage logic circuit drives a light-emitting diode (LED). The photo-optical signal from the LED is detected by a light-sensitive transistor in the high tension cell driving circuit, resulting in the release of an electrical current from a high-voltage power supply. This relatively inexpensive piece of equipment, which also generates a high-voltage AC signal and controls the coordinated switching between AC and DC outputs, has been used extensively for both electrofusion and electroporation of isolated plant protoplasts. The aim in designing this unit was to produce clean and consistent pulses with wide-ranging and flexible parameters, while meeting the high output demands of a Watts and King (1984) multiplate type of electrode, the advantages of which are discussed later. Some cell types cannot withstand the single large electrical pulse required to induce pore formation and may benefit from the application of small, multiple pulses. Unlike most commercial units, the equipment described by Jones et al. (1994) incorporates separate logic control over each pulse, thus enabling individual timing parameters to be assigned to each of a preset sequence of pulses. In addition to variation in pulse duration, ranging from 10 μsec to 3 msec, the delay between the firing of any two pulses (interpulse delay period) can be individually set between 0.28 and 5.9 sec by one of three continuously variable potentiometers. The ability to program DC pulse lengths independently has been used to introduce additional, short (probing) pulses into the cycle used to fuse isolated plant protoplasts. For example, the application of a 100-μsec DC pulse followed by three 10-μsec DC pulses resulted in fusion between isolated leaf protoplasts of *Porteresia coarctata* and protoplasts from cell suspensions of *Oryza sativa*, fusion between these protoplasts being unattainable with either single pulses or with multiple pulses of uniform duration (Finch et al., 1990).

Cell Alignment

The importance of the AC field in controlling electrofusion must not be ignored. As the AC field strength across a suspension of cells is increased, a voltage is reached at which the cells migrate and form pearl chains. To obtain the close membrane-to-membrane contact required for efficient cell or protoplast fusion, however, the dielectrophoretic force must be increased further to overcome the electrostatic and hydration barriers that cause repulsion between the membranes of adjacent cells (Rand, 1981). If the applied AC field is sufficient to achieve tight cell-to-cell compression, only small areas of localized membrane disruption are required to obtain cytoplasmic mixing. Consequently, the size of the DC pulse needed for fusion can be reached. Minimizing the level of membrane disruption thus assists in preserving the viability of fusion products. If the DC voltage and pulse length are maintained constant, the AC field can be used to regulate cell-to-cell compression, as observed with an inverted microscope, and, thereby, to control the fusion process to favor dimer formation. This ensures consistency between electrofusion experiments not obtained using duplicated voltage settings.

Many noncommercial electrofusion systems use a commercially built function generator as the AC source. These units, while providing the required signal range (0.5–2.0 MHz), usually have a maximum peak voltage output of only 15 V. Jones et al. (1994) increased the output of a Tcellner TCE 7702 function generator from 15 to 80 V by the addition of a parallel series of inexpensive, high-frequency pulse transformers (Radio Supplies Ltd., Corby, UK). To achieve the close membrane contact required to fuse small plant protoplasts, such as those of diameter about 20 μm, the output was further increased to 170 V by the insertion of a radio frequency amplifier (Sound Broadcast Services, 42 Grenville Road, London) between the function generator output and the pulse transformers. When amplifying the output, it is important that the internal resistance is matched to the electrofusion cell. Medium wave transmitters can provide a powerful AC output of the frequency range required for cell alignment. The sale of these transmitters is, however, illegal in some countries.

When reporting the electrical parameters used in a fusion experiment, it is important to remember that the electrical potential across the fusion chamber differs from the output voltage of the equipment used. If the resistance across the fusion chamber (load resistance) is lower than the internal resistance of the power supply, then the voltage across the fusion chamber may be significantly less than the open circuit voltage. The system developed by Jones et al. (1994) was linked to an oscilloscope to permit voltages at the electrode chamber to be monitored directly during fusion. As slight variations in resistance occur in cell samples, this facility aids reproducibility between experiments.

Electrode

The electrode assembly is of prime importance in terms of electrofusion efficiency, ease of use, and volumes of cell suspensions that can be processed. Certain conditions at the electrode assembly, such as the interelectrode gap and resistance, will influence the specification of the electrical equipment. It is convenient, in terms of both equipment design and operational safety, to use low-power circuitry to produce the required wave-forms. As stated earlier, however, the field intensity, E, is related to the applied voltage, V, and the interelectrode gap, d ($E = V/d$). Therefore, when the applied voltage is low, the interelectrode gap of the fusion chamber must be reduced to increase the field intensity to the required level. Several close proximity electrode assemblies have been described. Mischke et al. (1986) built an electrode by fixing four hypodermic needles, $800-1000$ μm apart, inside a 35 mm diameter Petri dish. The use of close electrodes limits the size of the chamber. Hence, only small volumes of cell suspensions can be processed during each electrofusion treatment. To circumvent this, Zachrison and Bornman (1984) designed three styles of fusion chamber on a flow-through principle. Although this made consecutive fusions easier, the volumes handled each time were still small (maximum 110 μL), and the only electrode of this type that allowed the fusion process to be monitored, using an inverted microscope, had a chamber volume limited to 49 μL. A further disadvantage of closely spaced electrodes is that cells tend to collect on the surfaces of the electrodes (Tempelaar and Jones, 1985). In particular, this is a problem when wire electrodes are used, since the nonuniform electrical field produced promotes the translational movement of cells toward the surfaces of the electrodes.

The electrofusion equipment developed by Jones et al. (1994) was designed to drive strong electrical fields across the relatively wide interelectrode gaps of a modified version of the Watts and King (1984) electrode. The electrode currently in use consists of seven, 2 cm wide, parallel nickel silver plates separated by perspex blocks each 2.7 mm in thickness, one of which is extended to form a handle. One mL aliquots of a suspension of parental protoplasts, mixed in the required ratio, are pipetted into the central nine wells of a 10×10 cm, 25-well replidish (Sterilin-Bibby, Stone, UK). The electrode is transferred to each well in succession. It is possible, by viewing the protoplasts using an inverted microscope, to monitor the fusion process. This electrode arrangement has certain advantages over the more commonly used 1-mm interelectrode gap fusion chambers (e.g., the Kruss multilamellar chamber) that are more suitable for the electromanipulation of animal cells, the latter being considerably smaller in size than isolated higher plant protoplasts. For example, on application of AC fields to induce cell alignment, widely spaced electrode plates favor mutual protoplast attraction rather than migration and adhesion of protoplasts to the electrodes (Tempelaar and Jones, 1985). This improves control over alignment and allows treatment of high-density protoplast suspensions. Furthermore, the ease of transfer of the electrode assembly between samples facilitates rapid processing of aliquots of parental protoplast mixtures (about 30–60 sec per well). Once fusion has been carried out, most of the electrofusion solution can be pipetted from above the settled protoplasts and replaced with culture medium. Electrofused protoplasts can be left undisturbed in situ for several hours to stabilize before further manipulations associated with culture.

User Safety

Most commercial electrofusion instruments have been designed for use with electrode assemblies with interelectrode gaps of 1 mm or less. Since the strength of the electrical field is dependent on the interelectrode gap, the output of many commercial instruments will not generate an effective electrical field across the 2.7-mm gap of a modified Watts and King (1984) type of electrode. This is also the case with most noncommercial equipment, where the alignment (AC) field is typically produced using a standard function generator. Although the high voltage levels required can be achieved in a number of ways, for example, by the use of a FCD device and a medium wave transmitter or by a purposely designed unit (Jones et al., 1994), an important consideration is operator safety. Jones et al., (1994) incorporated two forms of protection into their system. First, the output of the DC unit is monitored internally so should the logic circuits fail causing a prolonged DC current, the output is automatically terminated by a relay, and a warning sounder and indicator are activated. Second, the output socket to which the electrode is connected has a relay that will not complete the output circuit unless two disable pins are connected. Thus, to prevent accidental operator contact with the output voltage, a microswitch or similar device may be fitted to a protective guard so contact between these pins is only made when the equipment is in a safe operating enclosure.

SUMMARY

Several factors need to be considered when choosing equipment for electrofusion and electroporation. The convenience of purchasing an off-the-shelf system should be weighed against the requirements of the user in terms of the flexibility of the equipment, as well as cost. Noncommercial equipment must be constructed by persons with sufficient electronics expertise so that the instrument is built to a standard that will ensure safety of the user under all operating conditions.

References

Bates, G. W., Saunders, J. A., and Sowers, A. E. (1987). Electrofusion: Principles and applications. Pages 367–395. In Cell Fusion. Sowers, A. E., ed. Plenum Press, New York.

Blangero, C., and Teissié, J. (1985). Ionic modulation of electrically induced fusion of mammalian cells. J. Membr. Biol. **86:** 247–253.

Chang, D. C. (1989). Cell fusion and cell portion by pulsed radio-frequency electric fields. Pages 215–227. In Electroporation and electrofusion in cell biology. Neumann, E., Sowers, H. E. and Jordan, C. A., eds. Plenum Press, New York.

Chapel, M., Moutane, M. H., Ranty, R., Teissié, J., and Alibert, G. (1986). Viable somatic hybrids are obtained by direct current electrofusion of chemically aggregated plant protoplasts. FEBS Lett. **196:** 79–84.

Finch, R. P., Slamet, I. H., and Cocking, E. C. (1990). Production of heterokaryons by the fusion of mesophyll protoplasts of *Porteresia coarctata* and cell suspension-derived protoplasts of *Oryza sativa*: A new approach to somatic hybridization in rice. J. Plant Physiol. **136:** 592–598.

Fromm, M. E., Taylor, L. P., and Walbot, V. (1985). Expression of genes transferred into monocot and dicot plant cells by electroporation. Proc. Natl. Acad. Sci., USA. **82:** 5824–5828.

Glaser, R. W., Wagner, A., and Donath, E. (1986). Volume and ionic composition changes in erythrocytes after electric breakdown. Bioelectrochem. and Bioenerget. **16:** 455–467.

Joersbo, M., and Brunstedt, J. (1990). Direct gene transfer to plant protoplasts by electroporation by alternating, rectangular and exponentially decaying pulses. Plant Cell Rep. **8:** 701–705.

Jones, B., Antonova-Kosturkova, G., Vieira, M. L. C., Rech, E. L., Power, J. B., and Davey, M. R. (1993). High transient gene expression, with conserved viability, in electroporated protoplasts of *Glycine, Medicago* and *Stylosanthes* species. Plant Tissue Cult. **3:** 59–65.

Jones, B., Lynch, P. T., Handley, G. J., Malaure, R. S., Blackhall, N. W., Hammatt, N., Power, J. B., Cocking, E. C., and Davey, M. R. (1994). Equipment for the large-scale electromanipulation of plant protoplasts. BioTechniques **16:** 312–321.

Kinosita, K., and Tsong, T. Y. (1977a). Hemolysis of human erythrocytes by a transient electric field. Proc. Natl. Acad. Sci., USA. **74:** 1923–1927.

Kinosita, K., and Tsong, T. Y. (1977b). Formation and resealing of pores of controlled sizes in human erythrocyte membranes. Nature **268:** 438–441.

Kramer, D., Hsu, S., Miller, I., Riley, J., and Reporter, M. (1987). Circuit for the electromanipulation of plant protoplasts. Anal. Biochem. **163:** 464–469.

Lo, M. S. S., and Tsong, Y. T. (1989). Producing monoclonal antibodies by electrofusion. Pages 259–270. In Electroporation and electrofusion in cell biology. Neumann, E., Sowers, H. E., and Jordan, C. A., eds. Plenum Press, New York.

Lo, M. S. S., Tsong, T. Y., Conrad, M. K., Strittmatter, S. M., Hester, L. D., and Snyder, S. H. (1984). Monoclonal antibody production by receptor–mediated electrically induced cell fusion. Nature **310:** 794–796.

Mischke, S., Saunders, J. A., and Owens, L. (1986). A versatile low-cost apparatus for cell electrofusion and other electrophysiological treatments. J. Biochem. Biophys. Methods **13:** 65–75.

Neumann, E., Gerish, G., and Opatz, K., (1980). Cell fusion induced by high electric impulses applied to *Dictyostelium*. Naturwissenschaften **67:** 414–415.

Pietrzak, M., Shillito, R. D., Hohn, T., and Potrykus, I. (1986). Expression in plants of two bacterial antibiotic resistance genes after protoplast transformation with a new plant expression vector. Nucleic Acids Res. **14:** 5857–5868.

Pohl, H. A. (1951). The motion and precipitation of suspensoids in divergent electric fields. J. Appl. Phys. **22:** 869–871.

Pohl, H. A. (1958). Some effects of nonuniform electric fields on dielectrics. J. Appl. Phys. **29:** 1182–1189.

Pohl, H. A. (1978). Dielectrophoresis. Cambridge University Press, London.

Rand, R. P. (1981). Interacting phospholipid bilayers: Measured forces and induced structural changes. Ann. Rev. Biophys. Bioeng. **10:** 277–314.

Rech, E. L., Alves, E. S., and Davey, M. R. (1989). Electroporation: A circuit diagram and computer program for assessment of physical parameters on eucaryotic cells. Technique **1:** 125–129.

Saunders, J., Mathews, B. F., and Miller, P. D. (1989). Plant gene transfer using electrofusion and electroporation. Pages 343–354. In Electroporation and electrofusion in cell biology. Neumann, E., Sowers, H. E., and Jordan, C. A., eds. Plenum Press, New York.

Schwan, H. P. (1989). Dielectrophoresis and rotation of cells. Pages 3–21. In Electroporation and electrofusion in cell biology. Neumann, E., Sowers, H. E., and Jordan, C. A., eds. Plenum Press, New York.

Senda, M., Takeda, J., Abe, S., and Nakamura, T., (1979). Induction of cell fusion of plant protoplasts by electrical stimulation. Plant Cell Physiol. **20:** 1441–1443.

Sowers, A. E. (1984). Characterization of electric field-induced fusion in erythrocyte ghost membranes. J. Cell Biol. **99:** 1989–1996.

Sowers, A. E. (1985). Movement of a fluorescent lipid label from a labelled erythrocyte membrane to an unlabelled erythrocyte membrane following electric field-induced fusion. Biophys. J. **47:** 519–525.

Sowers, A. E. (1989). The mechanism of electroporation and electrofusion in erythrocyte membranes. Pages 229–256. In Electroporation and electrofusion in cell biology. Neumann, E., Sowers, H. E., and Jordan, C. A., eds. Plenum Press, New York.

Teissié, J., Knutson, V. P., Tsong, T. Y., and Lane, M. D. (1982). Electric pulse-induced fusion of 3T3 cells in monolayer culture. Science **216:** 537–538.

Teissié, J., and Rols, M. P. (1986). Fusion of mammalian cells in culture is obtained by creating the contact between cells after their electropermeabilization. Biochem. Biophys. Res. Commun. **140:** 258–266.

Tempelaar, M. J., and Jones, M. G. K. (1985). Fusion characteristics of plant protoplasts in electric fields. Planta **165:** 205–216.

Watts, J. W., and King, J. M. (1984). A simple method for the large-scale electrofusion and culture of plant protoplasts. Biosci. Rep. **4:** 335–342.

Weber, H., Forster, W., Berg, H., and Jacob, M. E. (1981). Parasexual hybridization of yeasts by electric field stimulated fusion of protoplasts. Curr. Genet. **4:** 165–166.

Zachrisson, A., and Bornman, C. H. (1984). Application of electric field fusion in plant tissue culture. Physiol. Plant. **61:** 314–320.

Zimmermann, U., and Scheurich, P. (1981). Fusion of *Avena sativa* mesophyll cell protoplasts by electrical breakdown. Biochim. Biophys. Acta. **641:** 160–165.

Zimmermann, U., and Vienken, J. (1982). Electric field-induced cell-to-cell fusion. J. Membrane Biol. **67:** 165–182.

Zimmermann, U., Vienken, J., Halfmann, J., and Emis, C. C. (1985). Electrofusion: A novel hybridization technique. Advances in Biotechnol. Processes **4:** 79–150.

CHAPTER 2

Electrically Stimulated Membrane Breakdown

T. Y. Tsong

ABSTRACT

The first sign of cell exposure to electrical pulses (strength in Kilovolts per Centimeter and duration in microseconds to milliseconds) is loss of the membrane permeation barrier against ions and small molecules. These permeability changes may be rapidly reversible or irreversible depending on the intensity and the width of the electrical pulses, as well as the composition of the suspending medium. After the rapid increase of the membrane permeability, many delayed effects of the electrical stimulation are observed. These slower secondary effects include membrane fusions, membrane bleb formation, endocytotic reactions, reorganization of the cytoskeletal network, and, in severe cases, lysis of the cells. Global membrane rupture and cell death are mainly due to these secondary effects. Cell death may be prevented by following certain protocols, the most crucial of which is to balance the osmotic pressure of the cytoplasmic fluid and the extracellular medium. Experiments show that electrical stimulation introduces pores of limited sizes in the plasma membrane. These pores can be resealed without losing the cytoplasmic macromolecular contents, and most cells will survive after pore resealing. Electroporation has found many applications in molecular biology, genetic engineering, agricultural research, and biotechnology.

INTRODUCTION

A cell interacts with an externally applied electrical field through its plasma membrane. The interaction is coulombic. Since the development of the membrane hypothesis of cells early this century, electrophysiology has centered around the study of the ion permeability and the electrical properties of the cell membrane (Cole, 1972; Jain and Wagner, 1980). The monograph by Cole documents early studies of electrical modification of ion conductivity of cell membranes and that by Tien (1974) of the lipid bilayer membrane (BLM). The generation of electrical impulses in neurons is the best known example of the reversible change of the Na^+ and K^+ conductivities of the plasma membrane (Kuffler and Nicholls, 1976). Artificially induced reversible breakdown of the ion permeation barrier was reported by Goldman (1943). He measured the voltage–current (V–I) characteristics of the membrane of *Chara australia* and found a phenomenon similar to the dielectrical breakdown of a cell membrane; a dramatic increase in conductance when the membrane was hyperpolarized beyond a critical value. When the voltage was reduced, the conductivity returned to normal values. Repetitive voltage scans did not change the V–I characteristics of the membrane, indicating that the membrane permeability changes were perfectly reversible.

Irreversible electrical breakdown of BLM was noted and has been studied in great detail to understand electrical characteristics of the lipid bilayer and the cell membrane (Huang et al., 1964; Tien, 1974). The dielectrical strength and other electrical parameters have been determined for BLM made of lipids from different sources and of different compositions (Tien, 1974). The dielectric strength of BLM for short electrical pulses (microseconds to milliseconds) is in the range 150–350 mV. This means that the critical breakdown field strength of the lipid bilayer, E_{crit}, is 300 to 700 kV/cm, assuming the thickness of the bilayer to be 5 nm. For cells in suspension, irreversible electrical breakdown of the plasma membrane was reported in the 1960s and 1970s (Pliquett, 1968; Sale and Hamilton, 1968; Coster and Zimmermann, 1975; Kinosita and Tsong, 1977). The main effects of the electrical stimulation of cells are leakage of ions, loss of cytoplasmic content, global changes in membrane morphology, and cell lysis (Pliquett, 1968; Sale and Hamilton, 1968; Kinosita and Tsong, 1977). When DC pulses were used, the critical breakdown membrane potential, $\Delta \psi_{crit}$, was found to be approximately 1 V. This translates into a E_{crit} of 2000 V/cm. Interest in the study of electroporation of cell membranes surged after the demonstration that electroporation could be used to load exogenous molecules such as drugs or DNA into living cells (Kinosita and Tsong, 1978; Tsong, 1987; Neumann et al., 1982; Wong and

Neumann, 1982). Electroporation has become a common laboratory technique for DNA transfection of cells (Neumann et al., 1989). This chapter summarizes work on the mechanistic studies of electroporation of cell membranes. Applications and other phenomena induced by pulsed electrical fields (PEF) are also discussed in chapters one and eight.

EFFECTS OF PULSED ELECTRICAL FIELDS ON CELL MEMBRANES

Field Effects and Thermal Effects

A pulsed electrical field can impose different kinds of effects on a chemical reaction in a homogeneous aqueous phase, due either to its electrical potential or to its current (Joule heating). Some of these effects are listed in Table 2-1 (Kinosita and Tsong, 1977; Tsong, 1983, 1990a). Ions or molecules with net charges will move in an electrical field (electrophoresis). Molecular dipoles or polarizable molecules will orient along the field line. An ion pair tends to dissociate under the influence of an electric field (Wien effect). Molecules with several conformational states of different molar electrical moments (M) will shift to a state with the highest electrical moment under an electrical field (electroconformational change). In an alternating electrical field (AC), these effects may take other forms. For example, if the field is nonuniform, a polarizable particle will move toward or away from regions of higher field intensity depending on the frequency of the AC field (dielectrophoresis). An AC field can also cause conformational oscillations of molecules by means of the electroconformational change. Temperature rise due to electrical current will change the equilibrium constant of a reaction according to the van't Hoff relationship (enthalpy). If the temperature change is abrupt, thermal expansion of the solvent will produce a shock wave that can also influence a chemical reaction. For reactions that occur in a cell membrane, all these effects are greatly amplified because the effective field experienced by these molecules is much greater than the applied field, as discussed later. In a cell, there are other field effects, such as electrocompression of the membrane. If a temperature gradient is established across a cell membrane because of nonuniform heating, water activity of the cytoplasm will be different from that of the external medium. Thermal osmosis effects would cause cell swelling. Colligative effects and thermal osmosis effects are, however, transient. A temperature gradient across a cell membrane with a micrometer diameter should dissipate in submicroseconds.

TABLE 2-1. Some Effects of an Electrical Field on Chemical Reaction.

	Effects Common to All Systems	Effects Specific to Cells in Suspension
Electrical field effects	Electrophoresis	Field-induced transmembrane potential
	Orientation of molecules	Amplification of common effects within cell membrane by $1.5 R_{cell}/d$ for a spherical cell
	Wien effect or ion pair dissociation	Large transmembrane current
For alternating field only	Electroconformational change $\Delta K/K = \Delta M \cdot \Delta E/RT$	Electrocompression of membrane
	Dielectrophoresis	Amplification of common effects within cell membrane
	Enforced conformational oscillating for energy and signal transductions	
Thermal effects current effects $\Delta T = i^2 r \Delta t/(4.18 C_P)$	Enthalpy effects $\Delta K/K = (\Delta H/RT^2)\Delta T$	Amplification of common effects due to transmembrane current
	Solvent expansion or shock wave $\Delta P = (\alpha/\kappa)\Delta T$	Thermal osmosis effects $\Delta \Pi = -(Q/\nu T)\Delta T$ Colligative effects $\Delta \Pi\text{-c}R\Delta T$

Symbols used: R_{cell}, radius of cell; ΔT, temperature change; r, specific resistivity of solution; C_P, specific heat capacity of solution; c, concentration of solute; K, equilibrium constant; ΔM, change in the molar electrical moment; ΔH, enthalpy of reaction; ΔP, pressure generated by shock wave; α, thermal expansion coefficient of solvent; κ, compressibility of solvent; Q, heat of transfer of water across cell membrane; d, membrane thickness; i, current; Δt, field exposure time; $\Delta \Pi$, change in osmotic pressure; R, gas constant; ΔK, change in equilibrium; ΔE, effective electrical field; T, Kelvin temperature; and ν, partial molar volume of solvent. See Kinosita and Tsong (1977) and Tsong (1983, 1990) for details.

Field-Induced Transmembrane Potential

When a spherical cell of radius R_{cell} suspended in a medium is exposed to an applied electrical field, there is a rapid accumulation of ions near the two loci facing the electrodes. The membrane will be hyperpolarized at the side close to the anode and depolarized at the side close to the cathode. The relaxation time, τ_{membr}, for the generation of the transmembrane potential, $\Delta \psi_{membr}$, depends on the membrane capacitance (C_{membr}), resistivities of the external medium (r_{ext}) and the cytoplasm (r_{int}), and the radius of the cell, as expressed in Equation 1:

$$\tau_{membr} = R_{cell} \, C_{membr} \, (r_{int} + r_{ext}/2)$$ (1)

The magnitude of $\Delta\psi_{membr}$ is described by Equation 2:

$$\Delta\psi_{membr} = 1.5 \, R_{cell} \, E_{appl} \cos\theta \, [1 - \exp(-t/\tau_{membr})]$$ (2)

in which E_{appl} and θ are the strength of the applied field and the angle between the field line and the normal from the center of the cell to a point of interest on the membrane surface. For biological cells of a micrometer diameter, τ_{membr} is less than 1 μ sec and the bracket in Equation 2 approaches unity in microseconds. For cells of larger diameter, τ_{membr} may have a value greater than 1 μ sec (Hibino et al., 1991). The maximal electrical potential across the plasma membrane (of thickness d) will be

$$\Delta\psi_{membr, max} = 1.5 \, R_{cell} \, E_{appl}$$ (3)

and the maximal effective electrical field will be $\Delta\psi_{membr, max}/d$.

Equations 1 and 2 have been derived from the Maxwell equations. The conditions required for these equations to be valid are that the spherical shell is made of thin, low conducting material and the conductivities of the internal and the external media are much higher than that of the shell. Experimentally, Equations 1–3 have been tested on lipid vesicles and cells by fluorescence imaging microscopy using potential sensitive fluorescence dyes (Gross et al., 1986; Ehrenberg et al., 1987; Kinosita et al., 1988; Hibino et al., 1991). Time resolution of sub microseconds has been achieved. Equation 3 has also been tested using the electroporation phenomenon (Kinosita et al., 1988; Hibino et al., 1991). In other experiments, single-component lipid vesicles of uniform size (95 ± 5 nm) were exposed to pulsed electrical fields. When the field intensity produced $\Delta\psi_{membr, max}$ of approximately 200 mV, the vesicles became permeable to radioactive tracers of sucrose, Rb^+ and Na^+ (Teissie and Tsong, 1981; El-Mashak and Tsong, 1985). This value agrees with the breakdown potential of BLM made of the same lipid, indicating that Equation 3 correctly calculated the breakdown potential of the lipid bilayer.

Because the generation of $\Delta\psi_{membr}$ by an applied field is not instantaneous, when an AC field is used, there will be a phase shift between the applied field and the induced transmembrane electrical field. Equations

1–3 can no longer be valid. For an AC field of the form shown in Equation 4, the Schwan equation (Equations 5 and 6) should apply (Schwan, 1983; Marszalek et al., 1990):

$$E_{appl} = E_{ac}^{\circ} \sin (2\pi f t) \tag{4}$$

$$\Delta \psi_{membr} = 1.5 \, R_{cell} \, E_{appl} \cos \theta / \left[1 + (2\pi f \tau_{membr})^2 \right]^{1/2} \tag{5}$$

$$\Delta \psi_{membr, max} = 1.5 \, R_{cell} \, E_{ac}^{\circ} / \left[1 + (2\pi f \tau_{membr})^2 \right]^{1/2} \tag{6}$$

E_{ac}° and f are the amplitude and the frequency of the AC field, respectively.

Equation 6 has been tested for myeloma cells by using a fluorescence probe, propidium iodide, to label DNA (Marszalek et al., 1990). When the $\Delta \psi_{membr, max}$ reached the membrane breakdown potential, the dye permeated into the cell, bound to DNA, and gave rise to two bright fluorescent bands near the two electrodes. The dye permeated the whole cell within 3 min. The AC field strength required to arrive at this membrane breakdown potential depended on the frequency as was predicted by Equation 6. Figure 2-1 summarizes the results of these experiments.

ELECTROPORATION OF HUMAN ERYTHROCYTE MEMBRANES

Formation of Aqueous Pores of Controlled Sizes

Work started from the finding that an exponentially decaying electrical pulse (initial strength of approximately 15 kV/cm and decay time of a few microseconds) when applied to human erythrocytes in isotonic saline led to hemolysis of the cells (Tsong and Kingsley, 1975). The cause of the hemolysis was not clear at the time, but thermal osmosis due to Joule heating was considered a possible cause. When lipid vesicles or *Bacillus subtlis* were treated with similar electrical pulses, many more molecular events were detected. Among these events were lipid phase transitions, vesicle swelling, membrane rupture, and vesicle fusions (Tsong 1974a, 1979b; Tsong and Kanehisa, 1977b). Membrane rupture was inferred from the turbidity measurement, and membrane fusion was deduced from the production of liposomes, when the initial sample consisted of small lipid vesicles.

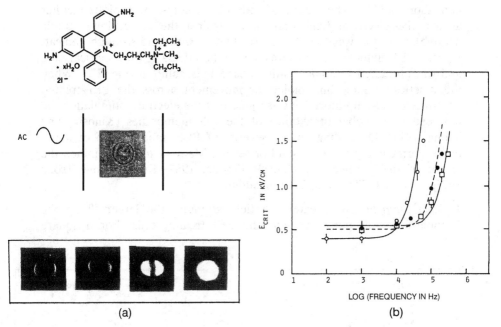

(a)   (b)

Figure 2-1. Frequency dependence of the critical field strength for electroporation by an alternating electrical field. (a) Visualization of membrane breakdown of
myeloma cell (Tib9) by the fluorescence changes of propidium iodide. The chemical formula of propidium iodide is shown in the upper part of Fig. 2-1(a).
Propidium iodide is weakly fluorescent in solution but becomes strongly fluorescent when it binds to DNA. The middle part of Fig. 2-1(a) shows a typical myeloma
cell under the light microscope. The relative position of the platinum electrodes
are indicated. The lower part of Fig. 2-1(a) gives four fluorescent photographs of a
myeloma cell taken at different times after the cell was electroporated by an AC
field (200 msec duration). Within 1–3 sec, two narrow, bright bands appeared at
the two loci facing the electrodes (the leftmost photo), indicating that the dye had
entered the cell at these two loci. The next three photos, from left to right, were
taken at 20 sec, 1 min, and 3 min, respectively, after electroporation. (b) Critical
AC amplitude (E_{crit}) for electroporation as a function of the AC frequency. Data
were obtained in three media of different resistivities, 52,600 Ω cm (O), 7050 Ω
cm (●), and 2380 Ω cm (□). The curves drawn through these data points were by
calculation (O) or by optimization (● and □) according to the Schwan Equation
(Equation 6). [Adapted from Marszalek et al. (1990)].

Kinosita and Tsong (1977a, 1977b; 1979) then systematically investigated
the phenomenon of the electrical pulse-induced membrane rupture using
human erythrocytes as a model system. To distinguish field effects and
thermal effects they did electrical stimulation of erythrocyte samples in
media consisting of the isotonic mixtures of NaCl and sucrose (Kinosita

and Tsong, 1977). The extent of red cell hemolysis was plotted either against the electrical field strength or against the power input (Joule heating). Cell lysis was found to correlate only with field strength, indicating that PEF-induced cell lysis was due to the effects of the electrical field and was unrelated to Joule heating of the bulk solution (Fig. 2-2). When the kinetics of ionic and molecular movement across the PEF-treated erythrocyte were monitored, it was found that the electrical stimulation did not lead to a global breakdown of the cell membranes (Kinosita and Tsong, 1977). Depending on the severity of PEF, modification of membrane permeability could be mild or severe. Many factors determined the extent of the permeability changes (Tsong, 1987; Kinosita and Tsong 1977a, 1977b, 1977c; 1978, 1979) including:

1. *Field Strength* The greater the field strength, the larger the probe molecules can permeate into the PEF-treated cells. For a square

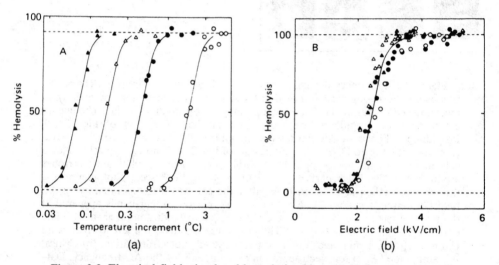

Figure 2-2. Electrical field stimulated hemolysis of human erythrocytes in media of different ionic strengths. An erythrocyte sample, suspended in a mixture of an isotonic NaCl (iso-NaCl) and an isotonic sucrose (iso-sucrose), was exposed to one square wave DC pulse of 20 μ sec with varied field strength. Joule heating was calculated from the current passing through the sample. Ten hours after electroporation, the extent of hemolysis was assayed by hemoglobin leakage. (a) Hemolysis is plotted against the temperature jump of the suspension in 100% iso-NaCl (○); 30% iso-NaCl plus 70% iso-sucrose (●); 10% iso-NaCl plus 90% iso-sucrose (△); 3% iso-NaCl plus 97% iso-sucrose (▲). The lack of correlation indicates that the hemolysis was not due to the Joule heating of the sample. (b) The sample set of data are plotted against the field strength of the DC pulses. A strong correlation is found between hemolysis and the field strength. [Adapted from Kinosita and Tsong, (1977a)].

electrical pulse of 20 μ sec, the critical field strength for membrane breakdown was approximately 2.0 kV/cm. This value translates into a critical breakdown potential of 1.0 V or a transmembrane field strength of 2000 kV/cm.

2. *Pulse duration* When square wave electrical pulses were used, the longer the pulse width, the larger the probe molecules can permeate the modified cell membranes. The critical breakdown potential did not, however, depend on pulse width for PEF longer than a few microseconds. This result indicates that the breakdown potential did not depend on the pulse width if the width is several times greater than τ_{membr}, which is in the sub microsecond for human erythrocytes. Pore expansion depended on pulse width and hence on the power input.

3. *Ionic strength* The severity of membrane modification was remarkably sensitive to the ionic strength of the suspending medium. The lower the ionic strength, the more severe the modification of the membrane permeability. For example, membranes of human erythrocytes suspended in isotonic saline when exposed to a single 3.7 kV/cm-20 μ sec PEF became freely permeable to Na^+ and K^+ but not to sucrose and hemoglobin. If the suspending medium was an isotonic mixture of 10% (w/v) NaCl/90% (w/v) sucrose, the membrane became permeable to the sucrose tracer. It remained impermeable to hemoglobin and other macromolecules.

4. *Temperature, divalent ions Ca^{++} or Mg^{++}, mercaptoethanol, sugars and alcohols* These also influenced the electrical stimulation of cell membrane permeability.

To interpret these observations, Kinosita and Tsong (1977b) proposed that PEF could implant aqueous pores of limited size on cell membranes. Their experiments using oligosaccharides of different mean radii as probes illustrated that the size of electropores could be controlled by adjusting the above electrical parameters and solvent conditions. Global rupture of cell membranes was not observed except in ill-controlled experiments, using excessively high field strength of long duration with a low ionic medium.

Colloidal Osmotic Hemolysis

The next step was to investigate the cause of hemolysis after electroporation of the cell membranes. Experiments showed that after the PEF treatment, cells began to swell in an isotonic medium. When the cell volume reached 155% of the normal volume, cell membranes ruptured and all cytoplasmic contents leaked out (Kinosita and Tsong, 1977a). Cell swelling was found to be due to the loss of the membrane permeation barrier to ions and small molecules: Electroporated membranes became semipermeable to cytoplasmic macromolecules (Kinosita and Tsong, 1977b). This resulted in an approximately 20–35 mOs (milli Osmoles) of excess osmotic pressure in the cytoplasm. The loss of the osmotic balance between the external medium and the cytoplasmic fluid drove ions and small molecules into the PEF-treated cells. The influx of water and the consequence of cell swelling eventually caused hemolysis of these membrane-perforated cells.

Resealing of Electropores without Loss of Cytoplasmic Macromolecules

To prevent colloidal hemolysis, 30 mOs of molecules of a size larger than electropores was added to the medium, and cell swelling was averted. Cells electrostimulated and suspended in a medium well balanced in osmotic pressure began to reseal. The resealing was exceedingly temperature sensitive (Kinosita and Tsong, 1977b). At 4°C cell swelling took days, but at 37°C it took only minutes (Fig. 2-3). The mechanisms of membrane resealing have not been studied in detail. The lack of interest in studying membrane resealing processes is because, in most cells, resealing is almost spontaneous. Erythrocytes are unusually fragile and require special treatment for membrane resealing. Most other cells do not lose much of their cytoplasmic contents after electroporation. Cells treated with PEF usually need a recovery period. After this period, they function and divide normally.

For erythrocytes, electroporated membranes must reseal completely or these cells will not survive in the circulation. The best criterion for complete membrane resealing is the recovery of the permeation barrier to K^+ or Rb^+. Serpersu et al., (1985) showed that small pores that allowed permeation of Rb^+, but not sucrose, took 15 min to reseal and pores that allowed sucrose permeation took more than 10 hr to reseal (Fig. 2-3).

Loading Exogenous Molecules into Living Cells

Kinosita and Tsong (1977b) showed that exogenous molecules could be loaded into electroporated erythrocytes that retained full hemoglobin content. In 1978 they demonstrated that mouse erythrocytes, which were loaded with radioactive sucrose and carefully resealed, survived in the circulation with a normal lifetime (Kinosita and Tsong, 1978; Tsong, 1987). Before these experiments, loading of drugs or enzymes by electroporation was achieved only in cell envelopes or erythrocyte ghosts (Zimmermann, 1982). For loading exogenous molecules into cell envelopes or ghosts, hypotonic shock is as effective as electroporation and the advantage of

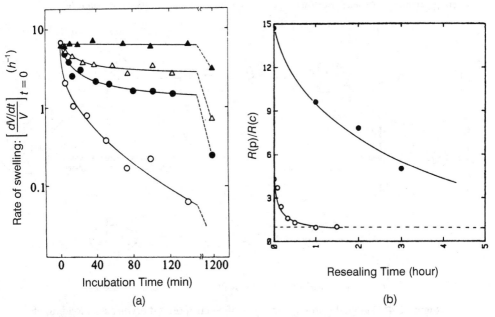

Figure 2-3. Time course of membrane resealing in electroporated human erythrocyte. (a) Temperature dependence. After erythrocyte was treated with a 3.7 kV/cm, 80 μ sec square wave DC pulse at 25°C, the sample was kept at different temperatures in a resealing medium containing osmotic pressure balancer, stachyose. At intervals, an aliquot was taken and added to iso-NaCl at 25°C. The initial rate of swelling was plotted against the time of incubation. The rate of swelling reflects membrane damage. Symbols indicate samples incubated at 3°C (▲), 17°C (△), 25°C (●), and 37°C (○). Hemolysis during these experiments was negligible. [Adapted from Kinosita and Tsong (1977b)]. Resealing of pores of different size. A sample (●) treated with a 4kV/cm-2 μ sec square wave DC pulse was permeable to Rb^+ but not sucrose. Its resealing against Rb^+ leak was complete in 1 hr. Another sample treated with a 3.7 kV/cm-20 μ sec pulse was initially permeable to sucrose. Its resealing against Rb^+ leak took more than 10 hr. Relative rate of Rb^+ permeation was plotted against time of incubation at 37°C in a resealing medium. [Adapted from Serpersu et al. (1987)].

electroporation is extremely limited. Figure 2-4 summarizes the colloidal hemolysis and entrapment of macromolecules into cell envelopes (upper path) and a protocol for drug loading and membrane resealing (lower path).

A great advance was made when Neumann et al., (1982) and Wong and Neumann (1982) succeeded in loading thymidine kinase (tk) gene into tk-deficient mouse L cells and demonstrated the expression of the loaded gene. Loading organelles and macromolecules into mammalian cells by the

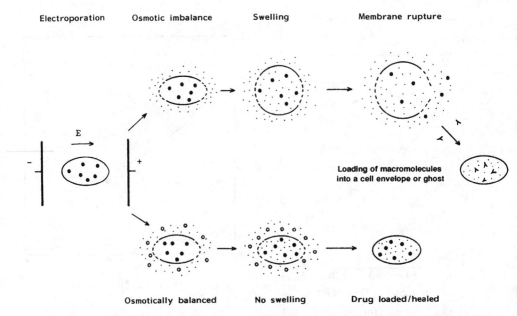

| Electroporation | Osmotic imbalance | Swelling | Membrane rupture |

Loading of macromolecules
into a cell envelope or ghost

Osmotically balanced   No swelling   Drug loaded/healed

Figure 2-4. Colloidal osmotic hemolysis of electroporated erythrocyte, loading of exogenous molecules, and resealing of electropores. In the figure, small dots denote drugs; small dark circles, cytoplasmic macromolecules; small open circles, inulin or oligosaccharide; Y-shaped molecules, antibodies or water-soluble enzymes. In the upper path, an electroporated erythrocyte swells due to the colloidal osmotic pressure of hemoglobin. When the membrane is ruptured, hemoglobin leaks out, and during the process of membrane resealing, antibodies or enzymes are loaded into the ghost. In the lower path, the colloidal osmotic pressure of hemoglobin is counterbalanced by inulin or oligosaccharide. Swelling of the electroperforated cells is prevented. A drug of small molar mass can be loaded and the cell membrane resealed completely. The final product is the drug-loaded intact erythrocyte. For most cells with stronger membranes than erythrocytes or with a cell wall, the upper path can be used to load macromolecules without total membrane rupture. Loss of some cytoplasmic content could happen, but cell lysis may be prevented.

PEF method has also been performed (Zimmermann, 1982; Neumann et al., 1982). Mechanistically, loading macromolecules or nanometer- and micrometer-size particles into living cells by the electrical stimulation is of particular interest. Uptake of macromolecules may involve endocytotic reactions, but evidence for such reactions remains scarce.

Pore Initiation and Expansion

Pore initiation may occur in lipid domains or at protein channels. In a lipid vesicle, pore initiation is likely to occur at existing defects of the crystalline lattice of the bilayer. Lattice defects exist at the boundary between a lipid domain in the gel state (crystalline state or solidlike state, S-state) and a domain in the liquid crystalline state (fluidlike state, F-state) (Tsong et al., 1977; Kanehisa and Tsong, 1978). In these interfacial regions, interactions among neighboring lipid molecules are weak. Lattice defects are found to facilitate permeation of ions and hydrophilic molecules (Tsong et al., 1977a, 1977b; Kanehisa and Tsong, 1978; Teissie and Tsong, 1981; El-Mashak and Tsong, 1985). Volume fluctuation is also the greatest in these regions (Ipsen et al., 1990) Molecular events forming a lattice defect include cis/trans isomerization of a methylene-to-methylene C–C bond, a reaction known to occur in 5 nsec (Gruenewald et al., 1981). Propagation of a hydrocarbon "kink" to its neighboring molecules is a highly cooperative process and takes place in many microseconds (Genz et al., 1986, Caffrey, 1989). The initiation of an electropore happens in the 1 μ sec time range (Serpersu et al., 1985). Pore initiation in the lipid bilayer requires a $\Delta\psi_{membr}$ of greather than 200 mV.

In a cell membrane, the opening/closing of protein channels is often controlled by transmembrane electrical potential. The gating potential of a typical ion channel is in the range of 50 mV, and the transit time of channels is in sub microseconds (Tsien et al., 1987). A PEF may force these channels to open before the breakdown of the lipid bilayer. Opening of protein channels is, however, unlikely to prevent a continuing rise of the $\Delta\psi_{membr}$, and pore initiation may occur in parallel in lipid domains and in protein channels.

As was pointed out, pore initiation depends only on the field strength if the pulse width is a few fold longer than τ_{membr}. After the pore initiation, there will be intense current across the pore that will generate local heating. In a lipid domain, heating increases molecular motion and prompts disordering of the lipid bilayer (Tsong et al., 1977a; Caffrey, 1989). These molecular events take place in a broad time range (microseconds to minutes). Concentric pores may also form due to a compression of the membrane or other mechanical effects. If a protein channel is punctured by an electrical potential, local heating due to excessive current may thermally denature the protein. Protein denaturation occur in milliseconds to seconds (Kim and Baldwin, 1990).

Experimentally, pore initiation and expansion have been measured by the generation of the transmembrane conductance in human erythrocyte, lipid vesicles, and BLM (Benz and Zimmermann, 1979; Kinosita and Tsong, 1979; Teissie and Tsong, 1981; Glaser et al., 1988). Time-resolved fluorescence imaging has also been used to follow kinetics of these events (Kinosita et al., 1988; Hibino et al., 1991). Consistent with the molecular dynamics discussed above, pore initiation occurs in the sub-microsecond time and pore expansion occurred in the 10 μ sec or longer time range.

Since pore initiation depends only on the field strength, it is purely electrical. In contrast, pore expansion depends on pulse width and is likely of thermal origin. These interpretations should be considered tentative, and there may be other explanations based on electromechanical phenomena (Powell et al., 1986).The above molecular events are best represented by kinetic schemes shown in Equations 7, 8, and 9 (Tsong, 1990);

$$A \rightleftharpoons B \rightarrow C \rightarrow C' \rightarrow B' \rightarrow A \qquad (7)$$

$$A \rightleftharpoons B \begin{array}{c} \nearrow C \rightarrow A_1 \\ \rightarrow D \rightarrow A_2 \\ \searrow E \rightarrow A_3 \end{array} \qquad (8)$$

$$P_{close} \rightleftharpoons P_{open} \rightleftharpoons P_{denatr} \begin{array}{c} \nearrow P_{irrev} \\ \downarrow \\ [O] \end{array} \qquad (9)$$

Equations 7 and 8 represent electroporation of a lipid vesicle or a lipid domain in a cell membrane. The A to B transition is the pore initiation step (sub microseconds to a few microseconds) and is considered reversible. If the transbilayer potential is not excessive enough to cause strong polarization of lipid molecules, molecular events that followed are best described by Equation 7. In this case, the B to C transition describes the expansion of pores, or the formation of hydrophilic pores. Pore expansion is relatively slow (microseconds to seconds). After the PEF is terminated, these pores reseal in milliseconds to seconds (Kinosita and Tsong, 1979). Since resealing takes place in the absence of an electrical field, the resealing of pores is not the exact reversal of the poration path; that is, the A to B to C transitions. If the applied field is high to induce a $\Delta \psi_{membr}$ of 1 V or greater, lipid molecules will be polarized. Electrostatic repulsions of polarized lipid molecules will cause fragmentation of the bilayer. The end result will be the production of smaller lipid vesicles than the original sample (Tsong, 1974a, b; Tsong and Kanehisa, 1977b).

In a cell membrane, electroporation through lipid domains should follow the path of Equation 7. Electroporation of protein channels will, however, follow Equation 9. In Equation 9, a protein channel is shown to be in several different conformational states—the closed, the open, the thermally denatured, and the irreversibly denatured. [O] denotes a protein that after being denatured is excised from the membrane; for example, by endocytotic process or by other cellular recycling mechanisms.

Resealing of Pores

Pores occurring in a lipid domain should reseal as represented in the C to C' to B' to A transitions in Equation 7. Once the plasma membrane of a cell is electroporated, however, the transmembrane electrical potential will rapidly dissipate. The after effects of this loss of a natural membrane potential could be the loss of lipid asymmetry and cytoskeletal network. When these events happen, a complete resealing will require restoration of the membrane permeation barrier, reestablishment of the membrane potential by ion pumping, reorganization of cytoskeletal network, and replenishment of cytoplasmic contents. Restoration of ion permeation barrier takes minutes to hours to complete (Kinosita and Tsong, 1977a, 1977b, 1978; Tsong, 1987; Serpersu et al., 1985). Other repair reactions of cellular faculty may take days (Neumann et al., 1989). Investigators working on electroporation or electrofusion of cells often find that the PEF-treated cells take a much longer period to restore their normal functioning than that of cells treated with chemicals such as polyethylene glycol (Neumann et al., 1989).

SOME RELATED PHENOMENA

Membrane Fusion

There are many reasons to believe that electrofusion of cell membranes is caused by the electroporation of cell membranes. First, the onset of field strength for electrofusion is identical to the onset of field strength for electroporation (Neumann et al., 1989). Second, after PEF treatment, cell membranes remain fusion competent (fusogenic state) for as long as the electropores are not completely resealed (Sowers and Lieber, 1986). Once these pores are resealed, the fusogenic state of cell membranes disappears. Membrane fusion is, however, a much more complex process. It not only requires the fusion of the lipid bilayer but also the intercellular weaving of the cytoskeletal network (Sowers, 1988; Neumann et al., 1989). The first documented observation of PEF-induced cell fusion was reported in 1974 by the author (Tsong, 1974a; 1990a). Fusion was thought to have been induced by Joule heating, which increases the fluidity of the lipid bilayer through the gel-to-liquid crystalline phase transition of lipid. Until now, the nature of the fusogenic state remains unclear. The increased fluidity of lipid domains after PEF treatment of cells remains a plausible explanation of the membrane fusion. Evidence suggests that a PEF uniformly modifies the plasma membrane to become fusogenic (Rols and Teissie, 1990). This observation contradicts many studies that indicate that electrical modification of cell membrane occurs preferentially at the two loci facing the electrodes, as predicted by Equations 3 and 5 (Kinosita et al., 1988; Marszalek et al., 1990; Hibino et al., 1991; see also Fig. 2-1). Electrofusion has been applied successfully to mammalian cells, bacteria, and plant protoplasts (Neumann et al., 1989). There are wide-spread applications of the electrofusion of cells in biomedical and agricultural research and biotechnology. Many of these applications are described elsewhere in this volume, in chapters 5, 6, 7, 9 and 10.

Membrane Bleb Formation

Pliquett (1968) reported that AC fields that exceeded a critical strength could produce severe morphological changes in *Oxytrichia* cells. Light microscopy revealed the formation of membrane blebs and endocytotic vesicles. Bleb formation has been investigated in greater detail by Gass and Chernomordik (1990). Satisfactory interpretations of these phenomena have not been offered. By considering the time sequence of molecular events discussed above, however, we can construct tentative mechanisms for future testing. After a cell is electroporated, there is a loss of the osmotic balance between the cytoplasmic fluid and the external medium. As depicted by rapid freeze electron microscopy, at the site of an electropore, the lipid bilayer appears to reform before the repair of the cytoskeletal tears (Chang and Reese, 1990). Since the overall volume of most cells are kept constant by a cell wall or a cytoskeletal meshwork, the colloidal swelling of a cell will preferentially take place through these newly formed lipid patches. Lipid patches are fluid and will balloon into bleblike structures. If this interpretation is correct, adding large molecules to the external medium to counterbalance the osmotic pressure of cytoplasmic fluid should prevent colloidal osmotic swelling of cells and hence the budding of membrane blebs. This appears to be the case (Chang and Reese, 1990). Formation of endocytotic vesicles may be a part of the cellular repairing process as was discussed and represented in the Equation 9.

Electrotransfection of Cells

Despite its popularity, mechanistic studies of electrotransfection of cells has remained an underexplored area. Some investigators favor electrophoresis as the driving force for DNA entry (Wong and Neumann, 1982; Neumann et al., 1982, 1989). A DNA molecule, being highly charged, will orient in an electrical field and move across the cell membrane by electrophoretic force. Many studies indicate that a membrane area covered by electropores is on the order of 0.01% (Kinosita and Tsong, 1977c; Sowers et al., 1986; Kinosita et al., 1988; Hibino et al., 1991). Experiments by Xie et al., (1990) showed that with a low DNA concentration (0.1 copy of DNA/cell), transfection of *Escherichia coli* by plasmid pBR322 could be as high as 0.7%. The probability of DNA encountering a pore should be smaller than 0.001% (0.1 × 0.01). Thus, electrophoresis of DNA through a pore is unlikely the mechanism for electrotransfection. Xie et al., (1990) further showed that the transfection efficiency correlated directly with the amount of DNA bound to the cell surface. DNA added many minutes after electroporation was also able to transfect the cells. In these few minutes, electroporated cell membranes remained in a competent state for DNA uptake. Surface binding and diffusion of DNA through electropores have been proposed to explain the above observations (Xie et al., 1990). Experiments by Chernomordik et al., (1990) indicate, however, that DNA taken by lipid vesicles after the PEF treatment is protected by lipid and is inaccessible to binding by ethidium bromide. Their results suggest that an endocytotic mechanism may be involved in the DNA uptake. Binding of DNA to the cell surface should also precede endocytosis in this case. Thus, both binding/diffusion and binding/endocytosis are plausible and further studies are warranted to see which of the two is the correct mechanism for the electrotransfection of DNA. Figure 2-5 compares several mechanisms for electrotransfection of a cell by a plasmid DNA.

SUMMARY

Although much attention has focused on the utilitarian aspect of electroporation and electrofusion, we should emphasize that "reversible" electrical modification of membrane permeability and induction of membrane fusions are a part of the natural processes of cells. Irreversible electroporation and electrofusion of cell membranes and the associated thermal denaturation of proteins and tissues are frequently encountered in clinical situations. Common electrical injuries and cardiac electrical shocks are two documented examples (Lee, 1990; O'Neill and Tung, 1991). A pulsed electrical field method has been used to insert genetically engineered membrane proteins into cell membranes (Mouneimne et al., 1990), activate membrane-associated enzymes (Teissie and Tsong, 1980; Serpersu and Tsong, 1983; Liu et al., 1990; Tsong, 1990b), and stimulate cell growth (Blank and Findl, 1987). Investigation of molecular mechanisms of these electrically stimulated cell membrane phenomena should provide us with essential data for the understanding of energy and signal transductions in cells and organisms. Understanding of molecular mechanisms should also

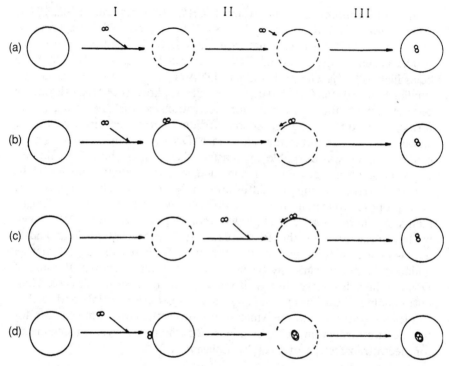

Figure 2-5. Different mechanisms of DNA transfection of cells by electroporation. In (a), after the electroporation of the cell (step I), circular and superhelical
plasmid DNA (indicated by ∞) diffuses into the cell through the bulk solution (step
II). Cell membranes reseal (step III). In (b), DNA binds to the cell surface, then
enters the cell through surface diffusion. In (c), electroporation is done before the
addition of DNA. DNA binds to the surface and diffuses into the cells through
stable pores. Because the size of pores is reduced during the incubation time (the
time between electroporation and the addition of DNA), the transfection efficiency
is much lower than that of (b). In (d), DNA is assumed to enter the cell by the
electrophoretic force of the applied field. When DNA enters, it carries with it a
piece of cell membrane or lipid bilayer and is prevented access by ethidium
bromide. Other mechanisms such as electrical field stimulated endocytosis for
DNA uptake are possible. [Adapted from Xie et al. (1990)].

help to design simpler and more reliable methods for applications in
biomedical, agricultural, and ecological research.

ACKNOWLEDGMENTS

The author thanks past and present colleagues for their contributions to
the project and C. J. Gross for help preparing the manuscript. This work
was supported by a grant from the U.S. Office of Naval Research.

References

Benz, R., and Zimmermann, U. (1979). Reversible electric breakdown of lipid bilayer membranes: A charge-pulse relaxation study. J. Membrane Biol. **48:** 181–204.

Blank, M., and Findl, E. (1987). Mechanistic approaches to interactions of electrical fields with living systems. Plenum Press. New York.

Caffrey, M. (1989). The study of lipid phase transition kinetics by time-resolved x-ray diffraction. Ann. Rev. Biophys. Chem. **18:** 159–186.

Chang, D. C., and Reese, T. S. (1990). Changes of membrane structure induced by electroporation as revealed by rapid-freezing electron microscopy. Biophys. J. **58:** 1–12.

Chernomordik, L. A., Sokolov, A. V., and Budker, V. G. (1990). Electrostimulated uptake of DNA by liposomes. Biochim. Biophys. Acta. 1024:179–183.

Cole, K. S. (1972). Membrane, ions and impulses. University of California Press, Berkeley.

Coster, H. G. L. and Zimmermann, U. (1975). The mechanism of electrical breakdown in the membranes of *Valonia utricularis*. J. Membrane Biol. **22:** 73–90.

Ehrenberg, B., Farkas, D. L., Fluhler, E. N., Lojewska, Z., and Loew, L. M. (1987). Membrane potential induced by external electrical field pulses can be followed with a potentiometric dye. Biophys. J. **51:** 833–837.

El-Mashak, E. M., and Tsong, T. Y. (1985). Ion selectivity of temperature-induced and electric field induced pores in dipalmitoylphosphatidylcholine vesicles. Biochemistry **24:** 2884–2888.

Gass, G. V., and Chernomordik, L. V. (1990). Reversible large-scale deformations in the membrane of electrically-treated cells: Electroinduced bleb formation. Biochim. Biophys. Acta **1023:** 1–11.

Genz, A., Holzwarth, J. F., and Tsong, T. Y. (1986). The influence of cholesterol on the main phase transition of unilamellar dipalmitoylphosphatidylcholine vesicles. Biophys. J. **50:** 1043–1051.

Glaser, R. W. Leikin, S. L., Chernomordik, L. V., Pastushenko, V. F., and Sokirko, A. I. (1988). Reversible electric breakdown of lipid bilayers: Formation and evolution of pores. Biochim. Biophys. Acta **940:** 275–287.

Goldman, D. E. (1943). Potential impedance and rectification in membranes. J. Gen Physiol. **27:** 37–50.

Gross, D., Loew, L. M. and Webb, W. W. (1986). Optical imaging of cell membrane potential changes induced by applied electric fields. Biophys. J. **50:** 339–348.

Gruenewald, B., Frisch, W. and Holzwarth, J. F. (1981). The kinetics of the formation of rotational isomers in the hydrophobic tail region of phospholipid bilayers. Biochim. Biophys. Acta **641:** 311–319.

Hibino, M., Shigemori, M., Itoh, H., Nagayama, K., and Kinosita, K. (1991). Membrane conductance of an electroporated cell analyzed by submicrosecond imaging of transmembrane potential. Biophys. J. **59:** 209–220.

Huang, C.-H., Wheeldon, L., and Thompson, T. E. (1964). The properties of lipid bilayer membranes separating two aqueous phases: Formation of a membrane of simple composition. J. Mol. Biol. **8:** 148–160.

Ipsen, J. H., Jorgensen, K., and Mouristen, O. G. (1990). Density fluctuations in saturated phospholipid bilayers increase as the acyl-chain length increases. Biophys. J. **58:** 1099–1107.

Jain, M. K., and Wagner, R. C. (1980). Introduction to Biological Membranes. John Wiley and Sons, New York.

Kanehisa, M. I., and Tsong, T. Y. (1978). Cluster model of lipid phase transitions with application to passive permeation of molecules and structure relaxations in lipid bilayers. J. Am. Chem. Soc. **100**: 424–432.

Kim, P. S., and Baldwin, R. L. (1990). Intermediates in the folding reactions of small proteins. Annu. Rev. Biochem **159**: 631–660.

Kinosita, K., Ashikava, I., Saita, N. Yosimura, H. Itoh, H. Nagayama, K., and Ikegami, A. (1988). Electroporation of cell membranes visualized under pulsed laser fluorescence microscope. Biophys. J. **53**: 1015–1019.

Kinosita, K., and Tsong, T. Y. (1977a). Hemolysis of human erythrocytes by a transient electric field. Proc. Natl. Acad. Sci. USA. **74**: 1923–1927.

Kinosita, K., and Tsong, T. Y. (1977b). Formation and resealing of pores of controlled sizes in human erythrocyte membrane. Nature **268**: 438–441.

Kinosita, K., and Tsong, T. Y. (1977c). Voltage induced pore formation and hemolysis of human erythrocyte membranes. Biochim. Biophys. Acta **471**: 227–242.

Kinosita, K., and Tsong, T. Y. (1978). Survival of sucrose-loaded erythrocytes in circulation. Nature **272**: 258–260.

Kinosita, K., and Tsong, T. Y. (1979). Voltage-induced conductance in human erythrocyte membranes. Biochim. Biophys. Acta **554**: 479–497.

Kuffler, S. W., and Nicholls, J. G. (1976). From Neuron to Brain. Sinauer Associates, Inc., Sunderland, Massachusetts.

Lee, R. C. (1990). Biophysical injury mechanisms in electrical shock victims. Proc. IEEE Eng. Med. Biol. Soc. Philadelphia. **12**: 1502–1504.

Liu, D.-S., Astumian, R. D., and Tsong, T. Y. (1990). Activation of Na^+ and K^+ pumping modes of Na,K-ATPase by an oscillating electric field. J. Biol. Chem. **265**: 7260–7267.

Marszalek, P., Liu, D.-S., and Tsong, T. Y. (1990). Schwan equation and transmembrane potential induced by alternating electric field. Biophys. J. **58**: 1053–1058.

Mouneimne, Y., Tosi, R.-F., Barhoumi, R., and Nicolau, C. (1990). Electroinsertion of full length recombinant CD4 into red cell membrane. Biochim. Biophys. Acta. **1027**: 53–58.

Neumann, E., Schaefer-Ridder, M., Wang, Y., and Hofschneider, P. H. (1982). Gene transfer into mouse lyoma cells by electroporation in high electric fields. EMBO J. **1**: 841–845.

Neumann, E. Sowers, A. E., and Jordan, C. A., eds. (1989) Electroporation and electrofusion in cell biology. Plenum Press, New York.

O'Neill, R. J., and Tung, L. (1991). A cell-attached patch clamp study of the electropermeabilization of amphibian cardiac cells. Biophys. J.

Pliquett, V. F. (1968). Das Verhalten von oxytrichiden unter einfluss des elektrischen, felds. Z. Biologie **116**: 10–22.

Powell, K. T., Derrik, E. G., and Weaver, J. C. (1986). A quantitative theory of reversible electrical breakdown. Bioelectrochem. Bioenerg. **15**: 243–255.

Rols, M.-P. and Teissie, J. (1990). Electropermeabilization of mammalian cells. Quantitative analysis of phenomenon. Biophys. J. **58** 1089–1098.

Sale, A. J. H., and Hamilton, W. A. (1968). Effects of high electric fields in microorganisms. III. Lysis of erythrocytes and protoplasts. Biochim. Biophys. Acta. **163**: 37–43.

Schwan, H. P. (1983). Biophysics of the interaction of electromagnetic energy with cells and membranes. Pages 213–231. In Biological Effects and Dosimetry of Nonionizing Radiation. Grandolfo, M., Michaelson, S. M., and Rindi, A. eds., Plenum Press, New York.

Serpersu, E. H., Kinosita, K. and Tsong, T. Y. (1985). Reversible and irreversible modification of membrane permeability by electric field. Biochim. Biophys. Acta. **812:** 779–785.

Serpersu, E. H., and Tsong, T. Y. (1983). Stimulation of Rb^+ pumping activity of Na, K-ATPase in human erythrocytes with an external electric field. J. Membrane Biol. **74:** 191–201.

Sowers, A. E. (1986). A long-lived fusogenic state is induced in erythrocyte ghosts by electric pulses. J. Cell Biol. **102:** 1358–1362.

Sowers, A. E. (1988). Fusion events and nonfusion content mixing events induced in erythrocyte ghosts by an electric pulse. Biophys. J. **54:** 619–626.

Sowers, A. E., and Lieber, M. L. (1986). Electropores in individual erythrocyte ghost: Diameter, lifetimes, numbers, and locations. FEBS Lett. **205:** 179–184.

Teissie, J., and Tsong, T. Y. (1980). Evidence of voltage induced channel opening in Na, K-ATPapse of human erythrocyte membranes. J. Membrane Biol. **55:** 133–140.

Teissie, J. and Tsong, T. Y. (1981). Electric field-induced transient pores in phospholipid bilayer vesicles. Biochemistry **20:** 1548–1554.

Tien, H. T. (1974). Bilayer Lipid Membranes (BLM): Theory and Practice. Marcel Dekker, New York.

Tsien, R. W., Hess, P., McClesky, E. W., and Rosenberg, R. L. (1987). Calcium channels: Mechanisms of selectivity, permeation, and block. Ann. Rev. Biophys. Biophys. Chem. **16:** 265–290.

Tsong, T. Y. (1974a). Temperature jump relaxation kinetics of aqueous suspensions of phospholipids and B. subtlis membranes. Federation Proceedings **33:** 1342.

Tsong, T. Y. (1974b). Kinetics of the crystalline-liquid phase transition of dimyristoyl L-α-lecithin bilayers. Proc. Natl. Acad. Sci. USA **71:** 2684–2688.

Tsong, T. Y. (1983). Voltage modulation of membrane permeability and energy utilization in cells. Bioscience Reports **3:** 487–505.

Tsong, T. Y. (1987). Electric modification of membrane permeability for drug loading into living cells. Methods in Enzymol. **149:** 248–259.

Tsong, T. Y. (1990a). On electroporation of cell membranes and some related phenomena. Bioelectrochem. Bioenerg. **24:** 271–295.

Tsong, T. Y. (1990b). Electrical modulation of membrane proteins: Enforced conformational oscillations and biological energy and signal transductions. Ann. Rev. Biophys. Biophys. Chem. **19:** 83–106.

Tsong, T. Y., Greenberg, M., and Kanehisa, M. I. (1977a). Anesthetic action on membrane lipids. Biochemistry **16:** 3115–3121.

Tsong, T. Y., and Kanehisa, M. I. (1977b). Relaxation phenomena in aqueous dispersions of synthetic lecithins. Biochemistry **16:** 2674–2680.

Tsong, T. Y., and Kingsley, E. (1975). Hemolysis of human erythrocyte induced by a rapid temperature jump. J. Biol. Chem. **250:** 786–789.

Wong, T.-K., and Neumann, E. (1982). Electric field mediated gene transfer. Biochem. Biophys. Res. Commun. **107:** 584–587.

Xie, T.-D., Sun, L. and Tsong, T. Y. (1990). Study of mechanisms of electric field-induced DNA transfection I. DNA entry by surface binding and diffusion through membrane pores. Biophys. J. **58:** 13–19.

Zimmermann, U. (1982). Electric field-mediated fusion and related electrical phenomena. Biochim. Biophys. Acta. **694:** 227–277.

CHAPTER 3

Cell Electrorotation

G. Fuhr

R. Hagedorn

ABSTRACT

In this chapter, we review the experimental and theoretical background of electrorotation and discuss the historical development of the method and its relationship to dielectrophoresis and impedance. This is followed by a brief description of the biological and technical applications of dielectric relaxation. The use of microfabricated chambers in new procedures for the manipulation or separation of dielectric particles suspended in liquids are also described. Summaries of the applications of electrorotation to the study of mammalian eggs and isolated chloroplasts provide an overview of the methods for measuring and interpreting the phenomenon.

INTRODUCTION

Several investigations into the behavior of suspended cells in high-frequency, alternating electric fields indicated that cells started to spin slowly under defined conditions (Teixera-Pinto et al., 1960; Füredi and Ohad, 1964; Pohl and Crane, 1971; Mischel and Lamprecht, 1980; Zimmermann et al., 1981). Holzapfel *et al.* (1982) correctly explained this phenomenon as the result of dipole-dipole interactions between neighboring cells. This led to the idea that isolated single cells could be induced to spin by the rotating electric field produced by a four-electrode arrangement (Arnold and Zimmermann, 1982a, 1982b). These workers observed that cells rotated much more slowly than the electric field vector and they found that cell rotation speed is dependent on the frequency of the field and proportional to the square of the field strength. The effect, later named 'electrorotation' has become a useful new method of dielectric spectroscopy for the study of individual living cells. It is related to dielectrophoresis and compliments impedance measurements (Glaser et al., 1985; Arnold and Zimmermann, 1988; Glaser and Fuhr, 1988).

Although the use of electrorotation in biology is relatively new, the application of rotating fields to the study of dielectrics has a long history, as have investigations of rotation phenomena in constant and alternating fields (Table 3-1).

THEORY

Physical Description

An electric field induces image charges at the interfaces between a liquid and a suspended body of different dielectric properties. If the field rotates, these charges travel synchonously, but lag behind due to charge relaxation in the body and surrounding medium. The permanent angular displacement between the rotating field vector and the image charges leads to a torque on the body and induces rotation. The torque is largest when the radian frequency of the field (ω) is close to the reciprocal of the dominant charge relaxation time (τ). At lower field frequencies, the induced charges are only slightly displaced and the resulting torque is small. At higher frequencies, very few image charges are induced, also leading to a small torque (Fig. 3-1). The torque causes the body to rotate, but, due to the viscous drag of the surrounding solution, the rotation speed is much less (typically 10^{-2} to 10^{-6}) than that of the field.

Cells can be driven either with or against the direction of field rotation (co-field and anti-field rotation; Fuhr and Kuzmin, 1986; Arnold et al., 1986; Arnold and Zimmermann, 1988). The direction of spin depends on the polarizability of the cell relative to the surrounding medium. For instance, in highly conductive solutions (above 0.5 S/m), only anti-field rotation of cells occurs (Fuhr et al., 1994). The various components of living cells (membrane, cell wall, cytoplasm or vacuoles) have different dielectric properties. Several charge relaxation processes are superimposed and each may dominate at a different frequency of the external field.

Cell rotation is inherently more complex than that of homogeneous, isotropic, dielectric particles. Living cells can, be characterized, however, by the use of rotating fields. The angular velocity of the cell is measured at

TABLE 3-1. Early Physical and Biological Papers Describing Rotational Effects in Electric Fields

Content	References
In Physics	
Experiments with mechanically produced rotating fields and a solar radiometer	Meyer (1877)
Theoretical work on conducting and dielectrical bodies in steady electrical fields	Hertz (1881)
Induced rotating electric fields	Arno (1892)
Experiments with dielectrics in constant, alternating, and quasirotating fields	Steinmetz (1892), Quincke (1896)
Criticism of Quincke's interpretation, presented in a quantitative theory	Heydweiller (1897, 1899), Schweidler (1897, 1907)
Induction of continuous rotation of dielectrics	Graetz (1900)
Fundamental and theoretical descriptions of the rotation effects of isotropic spheres	Lampa (1906)
Improved rotating field generation	Lang (1906)
Theory of rotation of permanent dipoles	Born (1920)
Experimental verification of Born's theory, measurements in four electrode chambers	Lertes (1921a, 1921b)
Nonspherical bodies in linear fields	Fürth (1924, 1927)
Experiments with constant, alternating, and rotating fields; detailed interpretation	Richardson (1927), Vedy (1931), Dänzer (1934), Krasny-Ergen (1937)
Development of dielectric Quincke motors	Sumoto (1956), Ataka and Namura (1959), Pickard (1961) Ogava (1961)
Rotating field measurements on crystals of cadmium sulfide	
Theory and experiments to Quincke rotation	Okano (1965), Secker and Scialom (1968)
Theory of birefringence of molecules in rotating fields	Funakoshi (1968)
Theoretical and experimental papers connected with dielectric Quincke motors	Coddington et al. (1970), Melcher and Taylor (1969); Secker and Belmont (1970); Simpson and Taylor (1971)
Theoretical background to calculation of rotational effects in constant and alternating fields	Melcher (1974)
Review on Quincke rotation	Jones (1984)
Electrostatic motors in planar silicon-etching techniques	Choi and Dunn (1971), Trimmer and Gabriel (1987), Tai et al. (1989), Kumar and Cho, (1990), Mehregany et al. (1990)
Theoretical analysis of dielectric induction motors in microstructural techniques	Choi and Dunn (1971), Bart et al. (1988), Fuhr et al. (1989a), Bart and Lang (1989), Fuhr et al. (1990a)
Force effects in dielectric liquids	Pickard (1965)
In Biology	
Theoretical background to interpret cell spectra	Sauer (1983), Hagedorn and Fuhr (1984), Fuhr (1985), Fuhr et al. (1985a, 1986, 1987b), Pastushenko et al. (1985), Sauer and Schlögl (1985), Wicher and Gündel (1985), Fuhr and Hagedorn (1987a), Fuhr and Kuzmin (1986), Schwan (1987), Pethig (1991a, b), Turcu and Lucaciu (1989)

TABLE 3-1. (*Continued*)

Content	References
Development of basic methodical principles	Pilwat and Zimmermann (1983), Pohl (1983a, 1983b, 1986), Fuhr et al. (1984), Arnold and Zimmermann (1987a, 1987b), Gimsa et al. (1987, 1988a, 1987b), Hözel (1988)
Reversible and irreversible polarization in rotating fields	Fuhr et al. (1989)
Yeast cells	Mischel and Lamprecht (1980), Mischel and Pohl (1983), Arnold et al. (1986, 1988a, 1988b), Geier et al. (1987), Hözel and Lamprecht (1987), Huang et al. (1992), Kaler and Johnston (1985)
Mammalian eggs	Arnold et al. (1987a, 1989), Fuhr et al. (1987a), Müller et al. (1988)
Isolated plant protoplasts	Arnold and Zimmermann (1982b), Fuhr et al. (1985a, 1987a), Glaser et al. (1983), Lovelace et al. (1984), Gimsa et al. (1985)
Protoplast fusion and rotation	Fuhr et al. (1987a, 1987c)
Isolated chloroplasts and thlakoid systems	Arnold et al. (1985), Fuhr et al. (1990b)
Erythrocytes	Mischel et al. (1982), Glaser et al. (1983), Mischel and Lamprecht (1983), Engel et al. (1988), Georgieva and Glaser (1988); Georgieva et al. (1989), Donath et al. (1990)
Platlets	Egger et al. (1988)
Thrombocytes	Egger et al. (1986)
Virus-cell interactions	Gimsa et al. (1989)
Lymphocytes	Ziervogel et al. (1986)
Cell separation	Fuhr et al. (1985b, 1985c)
Dielectric breakdown	Müller et al. (1986), Fuhr et al. (1989b)
Artificial particles	Hub et al. (1982), Küppers et al. (1983), Fuhr et al. (1985d), Wicher et al. (1986), Arnold et al. (1987b)
Erythroleukaemia cells	Wang et al. (1994)
Bacterial flagellar rotation	Washizu et al. (1993)
Review articles	Zimmermann (1982), Zimmermann and Arnold (1983), Arnold and Zimmermann (1984, 1988), Glaser et al. (1985), Glaser and Fuhr (1987, 1988), Arnold (1988), Fuhr and Glaser (1988), Fuhr and Hagedorn (1988)

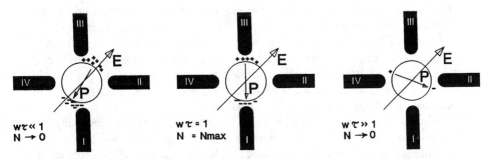

Figure 3-1. Orientation of the polarization vector (**P**) and field strength vector (**E**) at three characteristic frequencies (ω). The electrodes are marked I–IV and N is the torque acting on the particle. At $\omega\tau = 1$, a maximum torque is reached.

various frequencies of the driving field to yield a rotation spectrum (Arnold and Zimmermann, 1982b). The complexity of the spectrum increases with the number of dielectrically distinct layers and reflects the passive electrical properties of the cell components. Parameters such as conductivities and permittivities can be obtained by modelling experimentally obtained spectra.

Classification of Rotational Effects

If a rotating field polarizes a dielectric body, then electrorotation can generally be expected. Both the nature of the polarization process and the frequency of the field are significant. Following a proposal by Schwan (1957, 1987) the α, β, γ classification came into use. It is particularly applicable to the field of biology.

Polarization in the α frequency range is attributed to displacement of surface charges. Rotation at these low frequencies has been observed for artificial particles such as polystyrene spheres with carboxyl groups, electrets and liposomes (Hub et al., 1982; Küppers et al., 1983; Fuhr et al., 1985d; Wicher et al., 1986; Arnold et al., 1987b). Experimental difficulties arise when rotation is accompanied by electrophoretic movement or liquid streaming. Neither plant nor animal cells rotate strongly in the α range.

Living cells show two, well separated peaks in the β region of their rotation spectra. One occurs at kilohertz frequencies and reflects membrane charging/discharging processes. The other is in the megahertz frequency range and is due to polarization processes in the highly conductive cytoplasm and in the external solution; at these frequencies, the cell membrane offers little impedance to current flow and only a small potential is developed across it. This picture is only an approximation as, for instance, the electrical properties of the membrane may themselves be frequency-dependent. Relaxation processes of molecules within the bilayer may cause this, as may gating charges, mobile ions (Arnold and Zimmermann, 1988) or transport systems such as band III-protein in erythrocytes (Donath et al., 1990). Microscopic surface processes due to surface structures may also be of importance.

Finally, γ dispersion effects are due to relaxation processes involving the re-orientation of permanent molecular dipoles (e.g. water molecules). This phenomenon was investigated theoretically by Born in 1920 and experimentally verified by Lertes (1921a, 1921b) for several liquids. Living cells should also rotate in this frequency range (gigahertz).

A model spectrum, summarizing rotational behavior, is given in Figure 3-2.

Cell Models

The rotational behavior of living cells must be modeled and calculated if rotation spectra are to be interpreted. Cells are complex structures consisting of compartments of differing dielectric properties. All cells have a membrane which acts as a dielectric shell enclosing a highly conductive

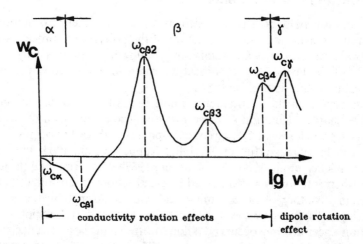

Figure 3-2. Idealized spectrum of a dielectric particle, demonstrating the frequency dependence and the superposition of rotational effects. ω_c is the angular velocity of the particle, $\omega_{c,\alpha..\gamma}$ are the field frequencies giving maximum rotation and α, β, γ are the dispersion ranges (for more information, see text).

TABLE 3-2. Cell Models Used for Calculation of Conductivity Rotation Effects

Cell Models	Number of Dielectrics	References
Isotropic to multishell sphere	$1 \cdots n$	Fuhr (1985), Fuhr et al. (1985a, 1986), Sauer and Schlögl (1985), Fuhr and Kuzmin (1986)
Isotropic to multishell cylinder	$1 \cdots n$	Fuhr (1985), Sauer and Schlögl (1985)
Isotropic to multishell ellipsoid	$1 \cdots n$	Sokirko (1992), Müller et al. (1993)

interior. Therefore, the simplest model is a single-shell body. Additional layers, such as a cell wall or other surface structures, necessitate the introduction of one or more extra shells into the model. Chloroplasts, nuclei and other organelles, with the exception of large vacuoles, are less polarized because the field is weakened by the conductive cytoplasm. These can, therefore, often be neglected. Multishelled bodies are well suited to modeling and were described in detail by Asami et al. (1984) and by Asmai and Irimajiri (1984). Cells, however, are extremely diverse in shape and few can be approximated as spheres, while analytical solutions to the Laplace-potential equation can only be found for special geometries. Numerical methods are usually required. Bodies with convexo/concave surfaces, such as erythrocytes, present particular problems (Stratton, 1941). Cell models, which have been used for the calculation of rotating field effects, are summarized in Table 3-2.

Theoretical Background

A particle exposed to a rotating electric field experiences a torque ($\underline{N}$) and the dielectrophoretic force ($\underline{F}$) which is caused by asymmetric polarization due to field inhomogeneities. These can be calculated by integrating the Maxwell stress tensor ($\overset{\leftrightarrow}{T}$) over a surface (S) which encloses the particle.

$$\underline{N} = \int_S (\underline{r} \times \overset{\leftrightarrow}{T}) \, d\underline{S} \qquad (1)$$

$$\underline{F} = \int_S (\overset{\leftrightarrow}{T} \cdot n) \, d\underline{S} \qquad (2)$$

where

$$T_{ik} = \varepsilon_o \varepsilon \left(E_i E_k - (\delta_{ik} E^2)/2 \right)$$

$\underline{r}$ is the radius vector; $\underline{n}$ is the outward normal vector; $\underline{E}$ is the field strength, ε is the permittivity and ε_0 is the permittivity of free space. Under certain conditions, this can be simplified to:

$$\underline{N} = (\underline{P}_{eff} \times \underline{E}_{ext}) \tag{3}$$

$$\underline{F} = (\underline{P}_{eff} \cdot \underline{\nabla})\underline{E}_{ext} \tag{4}$$

Here, $\underline{P}_{eff}$ is the induced effective dipole moment of the cell and $\underline{E}_{ext}$ is the external field strength. Rotating fields can be handled by the introduction of a complex function.

$$\underline{E}_{ext} = E_{oext} * \exp(j\omega t) \tag{5}$$

where E_{oext} is the amplitude of the external field, $j^2 = -1$ and ω is the angular frequency. The real part of E_{ext} is the x-component of the vector $\underline{E}_{ext}$ and the imaginary part is equal to the y-component. Thus, the induced dipole moment is also a complex function. The observable torque or dielectric force is the average value obtained by integrating over one revolution of the external field.

$$\langle \underline{N} \rangle = 0.5 * (\text{Im } p_{oeff}) * E_{oext} * \underline{k} \quad \text{or} \quad \langle N \rangle \propto \text{Im } p_{oeff} \tag{6}$$

$$\langle \underline{F}_x \rangle = 0.5 * (\text{Re } p_{oeff}) * \partial/\partial x(E_{oext}) \quad \text{or} \quad \langle F \rangle \propto \text{Re } p_{oeff} \tag{7}$$

Where $\underline{k}$ is the unit vector in the z-direction and p_{oeff} is the (complex) amplitude of the effective dipole moment.

As shown in Equations (6) and (7), the problem is reduced to the evaluation of the complex amplitude of the induced dipole moment. We shall demonstrate it for a single-shell sphere, the most often used cell model (Fig. 3-3), in a simplified manner considering only dipole moments.

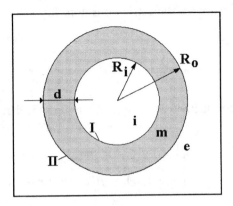

Figure 3-3. Single-shell sphere consisting of interior (i), membrane (m) and external medium (e). R_i and R_e denote the internal and external radii, d is the thickness of the membrane and the interfaces are marked I and II.

A solution to the Laplace equation must be found for the actual geometry. For a sphere, we obtain the following potentials (Φ) in each region.

$$\Phi_i = \underline{A} \cdot \underline{r} \qquad\qquad (r < R_i)$$

$$\Phi_m = \underline{B} \cdot \underline{r} + (\underline{C} \cdot \underline{r})/(4\pi\varepsilon_o r^3) \qquad (R_i < r < R_o)$$

$$\Phi_e = -\underline{E}_{\text{ext}} \cdot \underline{r} + (\underline{p}_{\text{eff}} \cdot \underline{r})/(4\pi\varepsilon_o\varepsilon_e r^3) \qquad (R_o < r) \qquad (8)$$

Where $\underline{A}$, $\underline{B}$ and $\underline{C}$ are complex vectors; $\varepsilon_o = 8.856 * 10^{-12}$ F/m.
Using the boundary conditions

$$\Phi_i = \Phi_m ; \Phi_m = \Phi_e \qquad (9)$$

and

$$\varepsilon_i^*(\partial\Phi_i/\partial r) = \varepsilon_m^*(\partial\Phi_m/\partial r) \qquad (10)$$

$$\varepsilon_m^*(\partial\Phi_m/\partial r) = \varepsilon_e^*(\partial\Phi_e/\partial r) \qquad (11)$$

and introducing complex dielectric constants for ε_i, ε_m and ε_e, respectively:

$$\varepsilon^*_{i,m,e} = \varepsilon_{i,m,e} - jG_{i,m,e}/(\varepsilon_o \omega) \tag{12}$$

where G is the conductivity, the equation system (8) can be solved and

$$P_{\text{eff}} = \frac{K\left[A1(\varepsilon_o \omega)^2 + jB1(\varepsilon_o \omega) + C1\right]}{\left(A2(\varepsilon_o \omega)^2 + jB2(\varepsilon_o \omega) + C2\right)} \tag{13}$$

where $K = 4\pi\varepsilon_o \varepsilon_e E_{\text{ext}} R_o^3$ and $d = R_0 - R_i$ (see Figure 3-3).

$$A1 = \varepsilon_m(\varepsilon_e - \varepsilon_i) + d(\varepsilon_i - \varepsilon_m)(2\varepsilon_m + \varepsilon_e)/R_o$$

$$A2 = -\varepsilon_m(\varepsilon_i + 2\varepsilon_e) - 2d(\varepsilon_i - \varepsilon_m)(\varepsilon_e - \varepsilon_m)/R_o$$

$$B1 = -G_m(\varepsilon_e - \varepsilon_i) - \varepsilon_m(G_e - G_i)$$

$$- d[(G_i - G_m)(2\varepsilon_m + \varepsilon_e) + (2G_m + G_e)(\varepsilon_i - \varepsilon_m)]/R_o$$

$$B2 = G_m(\varepsilon_i + 2\varepsilon_e) + \varepsilon_m(G_i + 2G_e)$$

$$+ 2d[(Ge - G_m)(\varepsilon_i - \varepsilon_m) + (G_i - G_m)(\varepsilon_e - \varepsilon_m)]/R_o$$

$$C1 = -G_m(G_e - G_i) - d(G_i - G_m)(2G_m + G_e)/R_o$$

$$C2 = G_m(G_i + 2G_e) + 2d(G_i - G_m)(G_e - G_m)/R_o$$

The real and imaginary parts of Equation (13) can be introduced into Equations (6) and (7). The torque leading to cell rotation can be written as

$$\langle N \rangle = 2\pi\varepsilon_o \varepsilon_e E^2_{\text{oext}} R_o^3$$

$$\times \left[\frac{B1\varepsilon_o \omega(A2\varepsilon_o^2\omega^2 + C2) - B2\varepsilon_o \omega(A1\varepsilon_o^2\omega^2 + C1)}{(A2\varepsilon_o^2\omega^2 + C2)^2 + (B2\varepsilon_o \omega)^2}\right] \tag{14}$$

This is a superposition of two relaxation processes. Under the special conditions of two, well-separated peaks, the characteristic frequencies (ω_{01} and ω_{02}) can be obtained.

$$\omega_{01} = C2/(\varepsilon_o B2) \quad \text{and} \quad \omega_{02} = -B2/(\varepsilon_o A2) \tag{15}$$

There is constant rotation since the electrically induced torque is exactly balanced by the frictional torque (N_f)

$$\langle N \rangle = N_f \tag{16}$$

where $N_f = 8\pi\eta R_o^3 \omega_c$ in the case of a sphere, η being the viscosity of the medium. Normally, the angular velocity of the cell (ω_c) is small compared to the field frequency (ω). Otherwise, ω in Equation (14) must be replaced by:

$$\omega_f = \omega - \omega_c \tag{17}$$

More details are given in Sauer and Schlögl (1985), Fuhr (1985) and Pastushenko et al. (1985).

As can be seen from Figure 3-4, cell rotation, in low conductivity solutions, occurs against the direction of field spin in the kilohertz range, while in the megahertz range the spins are in the same direction.

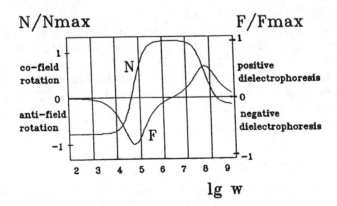

Figure 3-4. Dielectric force (F) and the torque (N) leading to rotation of a single-shell sphere. The curves have been calculated from the following parameters: $G_e = 10^{-2}$ S/m; $G_m = 10^{-7}$ S/m; $G_i = 0.1$ S/m; $e_e = 80$; $e_m = 8$; $e_i = 60$; $R_o = 25$ μm; $d = 6$ nm.

APPARATUS

Measuring Chamber

An example of an electrode configuration fabricated on a silicon microstructure is shown in Figure 3-5a (Fuhr et al., 1994).

The rotating fields can be created using various procedures (Fig. 3-6). However, field distributions in the central part may be non-uniform. Non-uniform fields create dielectrophoretic forces. These can be positive, attracting cells toward electrodes and other regions of high field strength, or negative, repelling cells into regions of low field strength (Pohl, 1978; Pethig, 1991a, 1991b; Jones, 1986). Also, streaming of the solution can occur due to self-stabilizing temperature gradients (Melcher, 1974) when frequencies near $G_o/(\varepsilon_o \varepsilon_e)$ are used. Thus, practical measuring chambers must ensure a large area of approximately uniform field distribution to avoid dielectrophoresis and suppress temperature gradients and streaming. Only in highly conductive solutions, where dielectrophoresis is always negative, are non-uniform fields helpful (Fuhr et al., 1994).

For significant cell spinning, there must be no friction other than the viscous drag of the suspending solution. Several possible strategies for achieving this are shown schematically in Figure 3-7.

Field Generator

The creation of rotating fields requires signals with frequency-independent phase-shifts. This may be achieved with analog devices (sine wave excitation, continuous rotation) or by digital generators (rectangular wave excitation, discontinuous rotation). The later are simpler and cheaper and experiments often use square-topped pulses (Pilwat and Zimmermann, 1983; Fuhr et al., 1984; Arnold, 1988; Gimsa et al., 1988b). Rotation spectra comparing continuously- and discontinuously-rotating excitation show deviations of less than 5% (Gimsa et al., 1988a). To achieve cell rotation speeds of approximately one revolution per second, field strengths of 4–10 kV/m must be applied.

Continuous Field Excitation

To create a field vector rotating with constant angular velocity, sine waves from a high-frequency generator are split and inverted to obtain, for example, four outputs with 90° phase shifts. The cost of controlling the phase shifts is a disadvantage, although sine waves allow the use of higher frequencies than do rectangular pulses (Arnold, 1988; Hölzel, 1988).

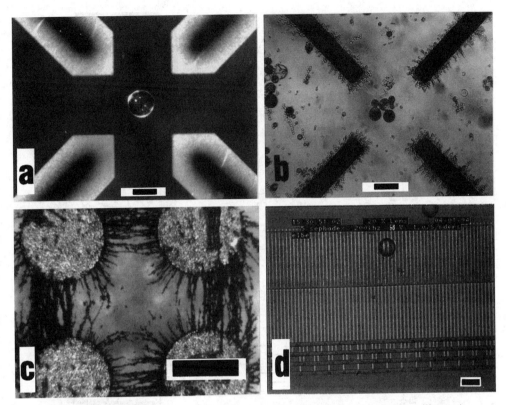

Figure 3-5. a) Four gold electrodes of 500 nm thickness processed on silicon. A rotating field centres a sephadex sphere of 70 μm diameter (bar = 50 μm). b) Four electrodes (gold on pyrex) are driven with square-topped pulses to separate two types of particle. Yeast cells show positive dielectrophoresis and are attracted to the electrodes, whereas cellulose spheres are focussed into the central part of the chamber (bar = 100 μm). c) Field lines in a four electrode chamber visualized with graphite particles (bar = 100 μm). d) Multielectrode chamber for the induction of linear motion of cells and micro-particles. The sphere on the upper part of the platinum electrodes is 70 μm diameter (bar = 70 μm).

Discontinuous Field Excitation

The output of a stabilized generator, tunable between 25 and 60 MHz, is connected to a chain of flip-flops which divide the primary frequency. The signal from a selected point in the chain is split to a divider and to an inverter followed by a divider. This combination of circuits produces exactly 90° phase-shifted signals independent of frequency. Depending on the number of electrodes, other phase angles can be used.

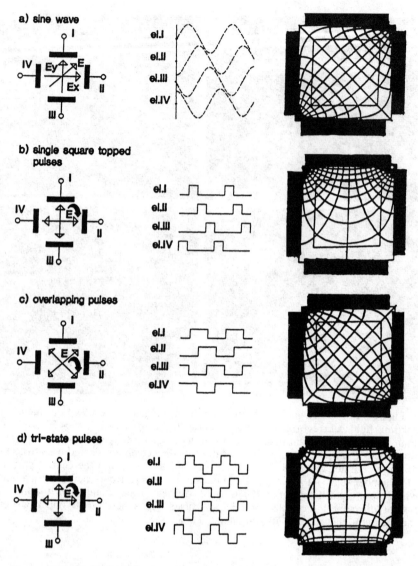

Figure 3-6. Field Distribution in a four electrode chamber under different regimes of driving voltage.

Measuring Technique

In the simplest case, particle rotation is measured as a function of frequency and the spectrum plotted semilogarithmically (Fig. 3-4). To measure the characteristic frequency of a peak, the compensation method

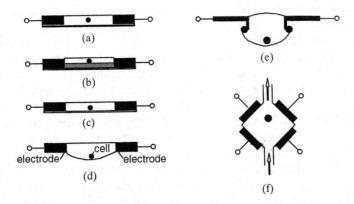

Figure 3-7. Various setups for the measurement of electrorotation (Glaser and Fuhr, 1987). a) Horizontal chamber for slowly sedimenting cells. b) Horizontal chamber with a density step. c) Chamber with bottom made of hydrophobic material to minimize friction. d) Chamber with a small hanging droplet. e) Chamber with a suspended droplet. f) Chamber generating lift by flow through vertical tubes.

of Arnold *et al.* (1988a, 1988b) can be used. This requires the application of two contra-rotating fields with a constant frequency ratio of $4:1$. If these fields act on a cell simultaneously or there is rapid, symmetric chopping between them, the resulting torque is zero when the geometric mean of the frequencies equals the characteristic frequency (ω_o). The strategy was extended by Gimsa (1987 to measure the whole spectrum. A rotating field of fixed frequency was used and the amplitude and frequency of the contra-rotating field were varied to obtain zero spin. The entire spectrum can be obtained purely from electrical signals which allows a high degree of automation. Technical details are given by Arnold and Zimmermann (1988) and Gimsa (1987).

INTERPRETATION OF DATA

Most cell spectra obtained in low conductivity solutions show anti-field rotation in the kilohertz range and co-field rotation at megahertz frequencies. Detailed analysis requires cell models. In principle there are two steps. A physical model, as described earlier, allows the measured frequencies and rotations to be transformed into physical constants (conductivities and dielectric constants). A second, more biophysical, model is necessary to convert these into ionic permeabilities, concentrations and membrane-characterizing parameters. This discussion is restricted to the first step, which is better understood.

Impedance measurements (Irimajiri et al., 1979) have revealed that living cells can seldom be regarded simply as membrane-enveloped bodies (single-shell spheres). They are better viewed as multilayered particles (cell walls and other surface structures) with more or less spherical internal compartments (vacuoles and organelles). This leads to multishell models. Each additional shell introduces at least three new parameters:—conductivity, permittivity and thickness. The behavior of multishell objects is of special relevance to plant protoplasts (tonoplast, cytoplasm, plasma membrane), mammalian egg cells (membrane, perivitelline space, zona pellucida) and isolated organelles such as chloroplasts (outer envelope membrane, intermembrane space, inner envelope membrane, interior). As an example, consider the electrorotational behavior of a thee-shell sphere as shown in Figure 3-8.

The solution of Equation 8, extended to the case of a three-shell sphere, requires 14 more or less unknown parameters (5 conductivities, 5 permittivities and 4 geometric factors). It might seem unrealistic to obtain these by data fitting. The following procedures can, however, render the problem tractable.

1. Theoretical predictions of both rotation and dielectrophoretic force should be made with all parameters varied in realistic ranges.

2. The rotational behavior of the cells should be measured at several different conductivities of the external solution.

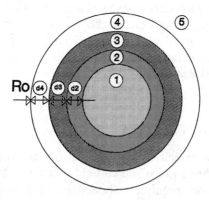

Figure 3-8. Three-shell model, consisting of four dielectrics representing (1) the interior; (2) inner membrane; (3) inter-membrane space; (4) outer membrane; and (5) the external medium.

3. The starting values for curve fitting should be determined by comparing actual and theoretical result for both rotation and dielectric force.

4. Additional measurements and the assumption of values for non-critical parameters will usually reduce the number of unknowns from 14 to between 5 and 7.

Figure 3-9 illustrates the behavior predicted, as parameters vary, for a three-shell body (corresponding to an isolated plant protoplast) whose properties are summarized in Table 3-3.

The dielectric constant of the cytoplasm (compartment 3, not shown in Fig. 3-9) has little effect on the spectrum. The external (compartment 5) and internal (compartment 1) conductivities are also not critical. The radius of the cell can be measured and the thicknesses of the membranes can be assumed to be 6 nm. The external conductivity is experimentally controlled. So 7 of the 14 parameters either are known, can be assumed or have no effect. This greatly simplifies the fitting process.

APPLICATIONS

Mammalian Eggs

Mammalian eggs are ideal for rotation studies as they are nearly spherical. Rotation spectra have been measured, in low conductivity media, for mouse eggs and for rabbit oocytes (Arnold et al., 1987a, 1989; Fuhr et al., 1987a; Müller et al., 1988). Many spectra exhibit three peaks, one anti-field in the kilohertz range and two co-field at megahertz frequencies (Fig. 3-10). An aid to understanding the rotational behavior of eggs is a two- or three-shell model in which the cytoplasm, membrane, perivitelline space and zona pellucida are represented. The processes underlying the spectral features can then be summarized:

1) The anti-field peak arises from membrane charging processes.

2) The first co-field peak reflects a superposition of effects in the zona pellucida and the perivitelline space.

3) The second co-field peak originates from polarization in the cytoplasm.

4) Increases in the external conductivity emphasize the three-peak nature of the spectrum.

The vitellus and zona were observed to rotate relative to each other (Arnold et al., 1987a, 1989; Fuhr et al., 1987a; Müller et al., 1988). This suggests that there is no rigid mechanical coupling between the two compartments and that the field-induced forces act at different points in

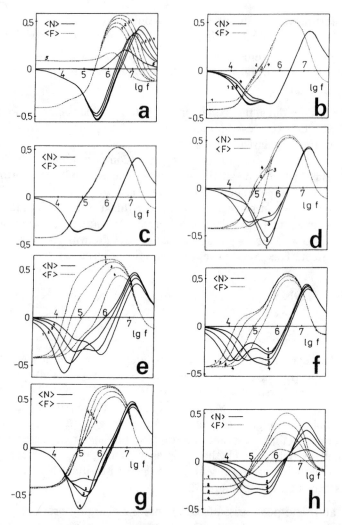

Figure 3-9. Theoretical frequency-dependence of the average torque $\langle N \rangle$ and the dielectrophoretic force $\langle F \rangle$, as predicted by a three-shell model with parameters given in Table 3-3 except for the following variations: a) G_1 (S/m): 1, 0.075; 2, 0.1; 3, 0.15; 4, 0.2; 5 is a simulation of a swollen plastid with highly conductive membranes ($G_2 = G_4 = 0.005$ S/m) and interior of low conductivity ($G_1 < 0.05$ S/m); b) G_2 (S/m): 1, 10^{-7}; 2, 10^{-6}; 3, $5 \cdot 10^{-6}$; 4, 10^{-5}; c) G_3 (S/m): nearly identical for values between 10^{-5} and 0.1; d) G_4 (S/m): 1, 10^{-7}; 2, 10^{-5}; 3, $5 \cdot 10^{-5}$; 4, $7 \cdot 10^{-5}$; e) G_5 (S/m): 1, 2.5; 2, 5: 3, 10; 4, 20; f) ε_2: 1, 40; 2, 20; 3, 8; 4, 4; g) ε_4: 1, 3; 2, 6; 3, 10; 4, 40; h) d_3 = intermembrane space (nm); 1, 992; 2, 492; 3, 242; 4, 8 (Fuhr et al., 1990b).

TABLE 3-3. Parameter Set Used for Calculation of the Rotational Behavior
of a Three-shell Sphere

Conductivities	Dielectric constants	Geometry
$G_1 = 0.1$ S/m	$\varepsilon_1 = 50$	$R_0 = 3\ \mu$m
$G_2 = 10^{-7}$ S/m	$\varepsilon_2 = 8$	$d_2 = 8$ nm
$G_3 = 0.01$ S/m	$\varepsilon_3 = 60$	$d_3 = 40$ nm
$G_4 = 5*10^{-5}$ S/m	$\varepsilon_4 = 3$	$d_4 = 8$ nm
$G_5 = 0.01$ S/m	$\varepsilon_5 = 80$	

the cell. Additionally, rotation measurements can be made on the isolated
zona pellucida and on the vitellus alone (Fig. 3-10). This facilitates the
calculation of cell parameters.

The anti-field peak of eggs isolated from superovulated and mated mice
is different from that of oocytes (Arnold et al., 1987a; Fuhr et al., 1987a).
Electrorotation, therefore, seems to be a promising technique for studies
into fertilization and early embryonic states.

Isolated Chloroplasts

Parallel measurements of electrorotation and dielectrophoresis have been
made on isolated chloroplasts (Fuhr et al., 1990b). After protoplast lysis,
plastids of *Avena sativa* L. were purified on a Percoll-density step-gradient

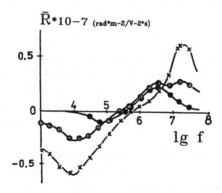

Figure 3-10. Rotation spectra of:—an intact oocyte (O), an oocyte without zona
pellucida (vitellus) (x) and an empty zona pellucida after electrical destruction of
the vitellus (·). The conductivity of the medium was 5 mS/m and the strength of
the rotating field was 8 kV/m.

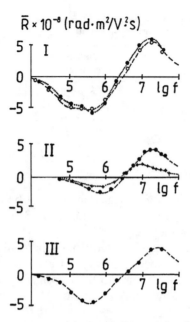

Figure 3-11. Rotation spectra of isolated chloroplasts (*Nicotiana tabacum L.*) purified on a Percol gradient (Fuhr et al., 1990b). Three types (I, II and III) of rotation spectra can be obtained from chloroplasts taken from the lowest interface. Only chloroplasts of type I have intact enveloping membrane systems.

(80%, 60%, 40%, 20%). The bottom layer (60%/80% contains a population of intact plastids. Three types of rotation spectra could, however, be distinguished (Fig. 3-11).

Dielectrophoretic motion was measured in the non-uniform field between a needle and a flat electrode (Fig. 3-12b). Plastids, in a droplet between the electrodes, were manipulated by movement of the microscope slide so that they started at the point x_1 (for positive dielectrophoresis) or at x_2 (for negative). The time for the plastids to travel the distance Δx was measured for a range of field frequencies. Velocity spectra are shown in Figure 3-12a.

Plotting the rotational and electrophoretic data in the Argand diagram gives semicircles in both the upper and lower halves (Fig. 3-13). This differs from the generally published Cole-Cole plots (Cole, 1972). The difference arises from the ability to measure rotation and dielectrophoresis on individual chloroplasts rather than phase-shifts in impedance methods (Fuhr and Glaser, 1988).

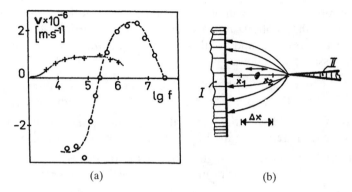

(a)   (b)

Figure 3-12. a) Velocity (v) as a function of field frequency (ω) for chloroplasts moving under dielectrophoretic force. ($\bigcirc$)—types I and III; ($+$)—type II. b) Electrode arrangement and field distribution used to measure dielectrophoretic motion in a non-uniform alternating field (Fuhr et al., 1990b).

The measurement of both rotation and dielectrophoresis enables chloroplast spectra to be analysed in more detail using multi-shell models and allows the investigation of the envelope membrane system and the average properties of the interior. Inner structures, such as thylakoid systems, do not influence rotation due to the field-weakening effect of the stroma.

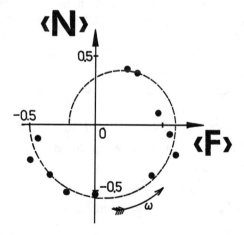

Figure 3-13. Cole-Cole plot of data obtained by electrorotation, $\langle N \rangle$, and dielectrophoresis, $\langle F \rangle$, measurements of individual chloroplasts. ω is the radian frequency of the field, $\langle N \rangle$ is torque/($2 \times$ maximum torque) and $\langle F \rangle$ is force/($2 \times$ maximum force). (Fuhr et al., 1990b).

The double envelope membrane becomes visible as the superposition of two peaks in the kilohertz range when there is an asymmetric distribution of electrical properties and the three types of spectra shown in Figure 3-11 can be classified as follows:

Type I: Chloroplasts with intact double envelope membrane systems, where the two membranes differ in conductivity or permittivity.

Type II: Chloroplasts with electrically equivalent membranes or with parameters differing by less than half an order of magnitude.

Type III: Chloroplasts with highly conducting (damaged) outer envelope membranes but with intact inner envelope membranes.

The ability to probe the outer envelope membrane should open up a range of possibilities for the characterization of isolated chloroplasts and the improvement of isolation procedures.

Microengineering

Highly miniaturized dielectric, side-drive motors have been developed using silicon microfabrication techniques (Choi and Dunn, 1971; Trimmer and Gabriel, 1987; Fan et al., 1989; Mehregany et al., 1990; Kumar and Cho, 1990). Most of them are electrostatic motors operating synchronously with the applied field and special geometries of rotor and stator electrodes are necessary.

An alternative is the dielectric relaxation motor (dielectric induction motor), which operates on the same principle as electrorotation (Choi and Dunn, 1971; Bart et al., 1988; Bart and Lang, 1989; Fuhr et al., 1989a, 1990a). This type of motor has certain advantages:

1. There is a variable torque-frequency characteristic which can be influenced by the architecture and by the combination of dielectrics in the rotor.

2. The torque is independent of the angular position of the rotor.

3. The rotor form is not dictated by the stator configuration.

4. The devices are extremely miniaturizable.

5. Operation up to 1000 Hz is possible.

6. The motor can be produced from silicon and its oxides.

Examples of architectures and rotation characteristics for three types of dielectric relaxation rotors are shown in Figure 3-14.

Induction and electrostatic motors are part of an extensive program to develop mechanical microsystems using the planar etching techniques of

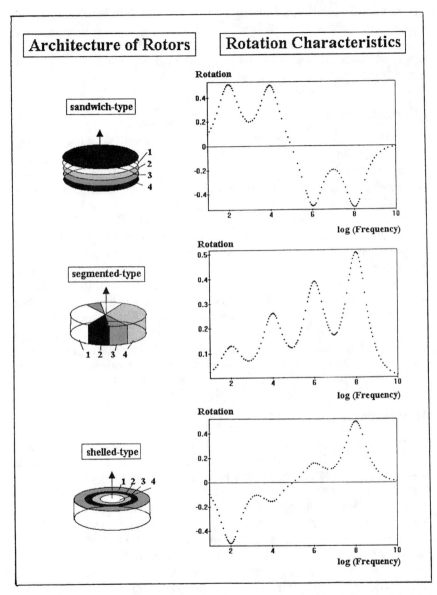

Figure 3-14. The characteristics of induction rotors can be influenced by the architecture and the combination of dielectric materials. The main rotor architectures shown here can be combined. Induction motors can be made extremely small (tens of microns) and have the advantage of a variable rotation vs. frequency characteristic.

silicon microfabrication technology. The aim is miniaturization and combination with electronics to produce complete measuring, sensing or actuating systems with chip dimensions for technical, biological, medical and other purposes. Living cells, inserted into silicon microstructures have found some practical uses (Washizu, 1989; Fuhr et al., 1990a).

Micromanipulation of Cells

Microstructuring can create electrode arrays which allow a linear varient of electrorotation (Figs. 3-5 and 3-15; Hagedorn et al., 1992). Such devices can be used to manipulate, position and separate cells. If the correct fields are applied, cells move linearly in the electrode gap and also over the electrodes.

There are several forces operating. There is a linear driving force which corresponds to the torque in the case of electrorotation. This arises from the phase-shift of the induced surface charges relative to the driving field, with the electrode track functioning as an 'unrolled' electrorotation chamber. The greatest force is developed at the frequency which would produce maximum rotation. In addition, there may be dielectrophoresis (which produces maximum force at a different frequency). If the dielectrophoresis is negative, cells will be forced into and levitated above the electrode gap, otherwise, they may roll on the surface of the electrodes but, under these conditions, friction increases dramatically and cell motion usually stops. Forces due to particle-particle interactions can also be important.

ADVANTAGES AND DRAWBACKS

Electrorotation is a noninvasive measuring technique. Advantages and disadvantages of the method are summarized in Table 3-4. Its main advantage is to allow observations to be made on single cells, paralleling microscopic or other observations. An advantage of using ultra-miniaturised electrode systems is the ease with which heat can be dissipated. This allows the application of strong a.c. fields (up to more than 50 kV/m) to highly conducting solutions such as cell culture media (e.g. DMEM) and physiological solutions (Fuhr et al., 1994). To characterise cells, a frequency range between several hertz and 100 MHz (in the future, possibly 1 GHz) is used and the technique is a new spectroscopic method.

As mentioned earlier, cell electrorotation requires models for analysis and for the calculation of physical parameters. Since cells vary in shape, analytical solutions to the Laplace equation often cannot be found. Equations and computer programs exist or are in preparation for homogeneous and multi-shelled spheres, infinitely long cylinders, ellipsoids and toroids.

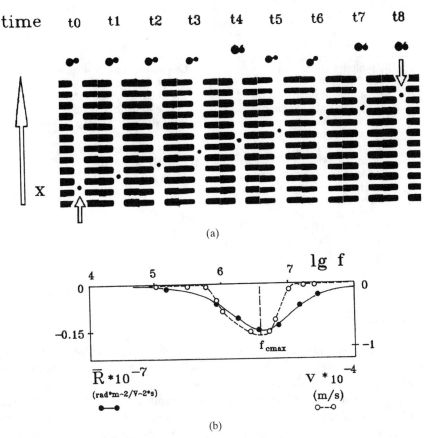

Figure 3-15. a) Multielectrode chamber (see also Fig. 3-5d) to induce linear motion of cells and particles. $t_{0.8}$ represent time intervals of one second each. The cellulose sphere is focused by negative dielectrophoresis into the electrode gap and moves along it in the opposite direction to the travelling field. The arrow, x, corresponds to 1 mm. b) Rotation (R) and linear velocity (v) of cellulose spheres as a function of frequency (f). Both rotation and velocity show a maximum at f_{cmax}. The fall off in linear velocity is caused by the force necessary to overcome the negative dielectrophoresis between each electrode pair.

Most cells can be approximated in one of these groups. Other geometries must be handled numerically.

In addition to cell rotation measurements, separation procedures have been developed and can be effected in multielectrode chambers by the application of alternating and rotating fields (Fuhr et al., 1990a, 1994). Such chambers allow the manipulation and separation of cells.

TABLE 3-4. Advantages and Disadvantages of Cell Electrorotation
and Linear cell motion

Advantages	Disadvantages
Suitable for single-cell investigations	Unsuitable for cell suspensions of high density
Measurement independent of the chamber volume	Conductivity of the external medium $G_i e < 0.2\,\text{S/m}$
High resolution of relaxation processes	Field strength $E > 1\,\text{kV/m}$
Simple and soft cell manipulation	Limited frequency range
Simple chamber construction	50 Hz–150 MHz (1 GHz)
Extremely miniaturizable	
Can be fabricated in micro-silicon technology	
Suitable for cell separation (size and also passive electrical properties)	

SUMMARY

This theoretical and experimental description of electrorotation gives only a general overview. A theoretical background exists which allows the simulation of the rotational behavior of particles, such as cells, with complex structures. Conductivities and permittivities can be calculated. Without doubt, electrorotation is a new and promising addition to passive electrical measuring techniques. It should be emphasised that, in every rotation chamber, there are four superimposed forces:—the torque leading to rotation, dielectrophoresis, gravitational/buoyancy forces and hydrodynamic forces.

In microfabricated chambers, alternating, rotating and traveling fields can be applied with low amplitudes, focusing particles to regions of high or low field strength or moving them unidirectionally. These can be used for cell investigation, manipulation and separation. The contactless manipulation of cells in chips with low-voltage, high-frequency fields seems to be of practical importance. Three dimensional manipulation in microfabricated labyrinths and experiments in small measuring chambers require solutions to many problems. The understanding of linear motion in travelling fields, rotation, dielectrophoresis and the trapping of cells in field cages (Schnelle et al., 1993; Fuhr et al., 1992) are, however, fundamental to the development of such microstructures.

ACKNOWLEDGMENTS

We are grateful to the Fraunhofer-Institut für Siliziumtechnologie (ISiT, Berlin) and especially to Dr. B. Wagner for the fabrication of microstructured chambers. We thank Dr. T. Müller for permission to use unpublished results and for many helpful discussions and Dr. S. G. Shirley for critically reading the manuscript. This work was supported by grant nos. 0310260A and 13MV03032 of BMFT (Germany).

References

Arno, R. (1892). Camo elettrico rotante per mezzo di difference di potenziali alternative. Atti Accad. Nuz. Lincei Rend. **1**: 139, 142.

Arnold, W. M. (1988). Analysis of optimum electro-rotation technique. Ferroelectrics **86**: 225, 244.

Arnold, W. M., Geier,. M., Wendt, B., and Zimmermann, U. (1966). The change in the electro-rotation of yeast cells effected by silver ions. Biochim. Biophys. Acta **889**: 35, 48.

Arnold, W. M., Schmutzler, R. K., Al-Hasani, S., Krebs, D., and Zimmermann, U. (1989). Differences in membrane properties between unfertilised and fertilised single rabbit oocytes demonstrated by electro-rotation. Comparison with cells from early embryos. Biochim. Biophys. Acta **979**: 142, 146.

Arnold, W. M., Schmutzler, R. K., Schmutzler, A. G., v.d. Ven, H., Al-Hasani, S., Krebs, D., and Zimmermann, U. (1987a). Electrorotation of mouse oocytes: Single cell measurements of zona-intact and zona-free cells and of the isolated zona pellucida. Biochim. Biophys. Acta **905**: 454, 464.

Arnold, W. M., Schwan, H. P., and Zimmermann, U. (1987b). Surface conductance and other properties of latex particles measured by electrorotation. J. Phys. Chem. **91**: 5093, 5098.

Arnold, W. M., Wendt, B., Zimmermann, U., and Korenstein, R. (1985). Rotation of a single swollen thylakoid vesicle in a rotating electric field: Electrical properties of the photosynthetic membrane and their modification by ionophores, lipophilic ions and pH. Biochim. Biophys. Acta **813**: 117, 131.

Arnold, W. M., and Zimmermann, U. (1982a). Rotation of an isolated cell in a rotating electric field. Naturwiss, **69**: 297.

Arnold, W. M., and Zimmermann, U. (1982b). Rotating field induced rotation and measurement of the membrane capacitance of single mesophyll cells of Avena sativa. Z. Naturforsch **37c**: 908, 915.

Arnold, W. M., and Zimmermann, U. (1984). Electric field-induced fusion and rotation of cells. Pages 389, 454. In Biological Membranes 5. Chapman, D., ed. Academic Press, London.

Arnold, W. M., and Zimmermann, U. (1987a). Verfahren und Vorrichtung zur Unterscheidung von in einem Medium befindlichen Partiken. Patentschrift DE 33 25 843 C2, Deutsches Patentamt, Applied 1983, Granted 1987.

Arnold, W. M., and Zimmermann, U. (1987b). Electrorotation: Experimental methods and results. Pages 514, 517. In Proceedings of 13th Annual NE Bioengineering Conference 1987. Forster, K. R., ed. IEEE, New York.

Arnold, W. M., and Zimmermann, U. (1988). Electrorotation: Development of a technique for dielectric measurements on individual cells and particles. J. Electrostatics 21: 151, 191.

Arnold, W. M., Zimmermann, U., Heiden, W., and Ahlers, J. (1988a). The influence of tetraphenyl borates (hydrophobic anions) on yeast cell electro-rotation. Biochim. Biophys. Acta 942: 96, 106.

Arnold, W. M., Zimmermann, U., Pauli, W., Benzing, M., Nierths, C., and Ahlers, J. (1988b). The comparative influence of substituted phenols (especially chlorophenols) on yeast cells assayed by electro-rotation and other methods. Biochim. Biophys. Acta 942: 83, 95.

Asami, K., and Irimajiri, A. (1984). Dielectric analysis of mitochondria isolated from rat liver. II. Intact mitochondria as simulated by a double-shell model. Biochim. Biophys. Acta 778: 570, 578.

Asami, K., Irimajiri, A., Hanai, T., Shiraishi, N., and Utsumi, K. (1984). Dielectric analysis of mitochondria isolated from rat liver. I. Swollen mitoplasts as simulated by a single-shell model. Biochim. Biophys. Acta 778: 559, 569.

Ataka, H., and Namura, S. (1959). On the rotation of dielectric in electrostatic field [in Japanese, English summary]. Denki Gakkai: J. Inst. Elec. Eng. Japan 79: 277, 281.

Bart, St. F., and Lang, J. H. (1989). An analysis of electroquasistatic induction micromotors. Sensors and Actuators 20: 97, 106.

Bat, St. F., Lober, Th. A., Howe, R. T., Lang, J. H., and Schlecht, M. F. (1988). Design consideration for micromachined electric actuators. Sensors and Actuators 14: 269, 292.

Born, U. (1920). Über die Beweglichkeit der elektrolytischen Ionen. Z. d. Phys. 1: 221, 249.

Choi, S. D., and Dunn, D. A. (1971. A surface-charge induction motor. Proc. IEEE 59: 737, 748.

Coddington, T., Pollard, A. F., and House, H. (1970). Operation of a dielectric motor with a low conductivity liquid. J. Phys. D: Appl. Phys. 3: 1212, 1218.

Cole, K. S. (1972). Membranes, ions and impulses. University of California Press, Berkeley.

Dänzer, H. (1934). Über das Verhalten biologischer Körper im Hochfrequenzfeld. Ann. d. Phys. 20: 463, 480.

Donath, E., Egger, M., and Pastushenko, V. Ph. (1990). Dielectric behaviour of the anion exchange protein of human red blood cells. Theoretical analysis and comparison to electrorotation data. Bioelectrochem. Bioenerg. 23: 337, 360.

Egger, M., Donath, E., Spangenberg, P., Bimmler, M., Glaser, R., and Till, U. (1988). Human platelet electrorotation change induced by activation: Inducer specifity and correlation to serotonin release. Biochim. Biophys. Acta 972: 265, 276.

Egger, M., Donath, E., Ziemer, S., and Glaser, R. (1986). Electrorotation: A new method for investigating membrane events during thrombocyte activation. Influence of drugs and osmotic pressure. Biochim. Biophys. Acta 861: 122, 130.

Engel, J., Donath, E., and Gimsa, J. (1988). Electrorotation of red cells after Electroporation. Stud. Biophys. 125: 53, 62.

Fan, L. Sh. Tai, Y.-Ch., and Muller, R. S. (1989). IC-processed electrostatic micromotors. Sensors and Actuators 26: 41, 47.

Fuhr, G. (1985). Über die Rotation dielektrischer Körper in rotierenden Feldern. Dissertation, Humboldt-Universität, Berlin.

Fuhr, G., Arnold, W. M., Hagedorn, R., Müller, T., Benecke, W., Wagner, B., and Zimmermann, U. (1992). Levitation, holding and rotation of cells within traps made by high frequency fields. Biochim. Biophys. Acta **1108**: 215, 223.

Fuhr, G., Geissler, F., Müller, T., Hagedorn, R., and Torner, H. (1987a). Differences in the rotation spectra of mouse oocytes and zygotes. Biochim. Biophys. Acta **930**: 65, 71.

Fuhr, G., Gimsa, J., and Glaser, R. (1985a). Interpretation of electrorotation of protoplasts. I. Theoretical consideration. Stud. Biophys. **108**: 149, 164.

Fuhr, G., and Glaser, R. (1988). Electrorotation: A new method of dielectric spectroscopy. Stud. Biophys., **127**: 11, 18.

Fuhr, G., Glaser, R., Hagedorn, R. (1986). Rotation of dielectrics in a rotating electric high-frequency field. Model experiments and theoretical explanation of the rotation effect of living cells. Biophys. J. **49**: 395, 402.

Fuhr, G., and Hagedorn, R. (1987a). Rotating-field-induced membrane potentials and practical applications. Stud. Biophys. **119**: 97, 101.

Fuhr, G., and Hagedorn, R. (1988). Grundlagen der Elektrorotation. *In* Colloquia Pflanzenphysiologie. Göring, H., and Hoffmann, P., eds. Humboldt-Universität, Berlin.

Fuhr, G., Hagedorn, R., Glaser, R., and Gimsa, J. (1989a). Dielektrische Motoren. Elektrie **43**: 45, 50.

Fuhr, G., Hagedorn, R., Glaser, R., Gimsa, J., and Müller, T. (1987b). Membrane potentials induced by external rotating fields. J. Bioelectricity **6**: 49, 69.

Fuhr, G., Hagedorn, R., and Gimsa, J. (1990a). Rotational behaviour of living cells with reference to micro-motors. Pages 832, 837. *In* Micro System Technologies 90. Reichel, H., ed., Springer-Verlag, Berlin-Heidelberg.

Fuhr, G., Hagedorn, R., and Göring, H. (1984). Cell rotation in a discontinuous field of a 4-electrode chamber. Stud. Biophys. **102**: 221, 227.

Fuhr, G., Hagedorn, R., and Göring, H. (1985b). Separation of different cell types by rotating electric fields. Plant and Cell Physiol. **26**: 1527, 1531.

Fuhr, G., Hagedorn, R., and Müller, T. (1985c). Cell separation by using rotating electric fields. Stud. Biophys. **107**: 23, 28.

Fuhr, G., Hagedorn, R., and Müller, T. (1985d). Simulation of the rotational behaviour of single cells by macroscopic spheres. Stud. Biophys. **107**: 109, 116.

Fuhr, G., and Kuzmin, P. (1986). Behaviour of cells in rotating electric fields with account to surface charges and cell structures. Biophys. J. **50**: 789, 795.

Fuhr, G., Müller, T., and Hagedorn, R. (1989b). Reversible and irreversible rotating field-induced membrane modifications. Biochim. Biophys. Acta **980**: 1, 8.

Fuhr, G., Müller, T., Wagner, A. and Donath, E. (1987c). Electrorotation of oat protoplasts before and after fusion. Plant and Cell Physiol. **28**: 549, 555.

Fuhr, G., Roesch, P., Müller, T., Dressler, V., and Göring, H. (1990b). Dielectric spectroscopy of chloroplasts isolated from higher plants. Plant and Cell Physiol. **31**: 975, 985.

Fuhr, G., Glasser, H., Müller, T., and Schnelle, Th. (1994). Cell manipulation and cultivation under a.c. electric field influence in highly conductive culture media. Biochim. Biophys. Acta **1201**: 353, 360.

Funakoshi, H. (1968). Theory of the birefringence caused by the orientation of particles in the rotating electric field. Ann. Inst. Stat. Math. Supp. V: 45, 66.

Füredi, A. A., and Ohad, I. (1964). Behaviour of human erythrocytes in high-frequency electric fields and its relation to their age. Biochim. Biophys. Acta **79**: 1, 8.

Fürth, R. (1924). Eine neue Methode zur Bestimmung der Dielektrizitäts-konstanten guter Leiter. Z. Phys. **22**: 98, 108.

Fürth, R. (1927). Die absolute Bestimmung von Dielektrizitätskonstanten mit Ellipsoidmethode. Z. Phys. **44**: 256, 260.

Geier, B. M., Wendt, B., Arnold, W. M., and Zimmermann, U. (1987). The effect of mercuric salts on the electro-rotation of yeast cells and comparison with a theoretical model. Biochim. Biophys. Acta **900**: 45, 55.

Georgieva, R., Donath, E., Gimsa, J., Löwe, U., and Glaser, R. (1989). AC-field-induced KCl leakage from human red cells at low ionic strengths: Implications for electrorotation measurements. Bioelectrochem. Bioenerg. **22**: 255, 270.

Georgieva, R., and Glaser, R. (1988). Electrorotation of lidocaine-treated human erythrocytes. Pages 263, 266. *In* Electromagnetic Fields and Biomembranes. Markov, M., and Blank, M., eds. Plenum Press, New York.

Gimsa, J. (1987). Elektrorotation: Technische Voraussetzungen und biophysikalische Aussagemöglichkeiten. Dissertation. Humboldt-Universität, Berlin.

Gimsa, J., Donath, E., and Glaser, R. (1987). Electrorotation: Influence of pulse shape and internal membrane systems on rotation spectra. Pages 173, 177. *In* Electromagnetic Fields and Biomembranes. Blank, M., and Markov, M. eds. Plenum Press, New York.

Gimsa, J., Donath, E., and Glaser, R. (1988a). Evaluation of data of simple cells by electrorotation using square-topped fields. Bioelectrochem. Bioenerg. **19**: 389, 396.

Gimsa, J., Fuhr, G., and Glaser, R. (1985). Interpretation of electrorotation of protoplasts. II. Interpretation of experiments. Stud. Biophys. **109**: 5, 14.

Gimsa, J., Glaser, R., and Fuhr, G. (1988b). Remarks on the field distribution in four electrode chambers for electrorotational measurements. Stud. Biophys. **125**: 71, 76.

Gimsa, J., Pritzen, C., and Donath, E. (1989). Characterization of virus-cell-interaction by electrorotation. Stud. Biophys. **130**: 123, 131.

Glaser, R., and Fuhr, G. (1987). Electrorotation of single cells: A new method of assessment of membrane properties. Pages 271, 290. *In* Electric Double Layers in Biology. Blank, M., ed. Plenum Press, New York.

Glaser, R., and Fuhr, G. (1988). Electrorotation: The spin of cells in rotating high frequency electric fields. Pages 271, 290. *In* Mechanistic Approaches to Interactions of Electric and Electromagnetic Fields with Living Systems. Blank, M., and Findl, E., eds., Plenum Press, New York.

Glaser, R., Fuhr, G., and Gimsa, J. (1983). Rotation of erythrocytes, plant cells and protoplasts in an outside rotating electric field. Stud. Biophys. **96**: 11, 20.

Glaser, R., Fuhr, G., Gimsa, J., Hagedorn, R. (1985). Electrorotation: Capabilities and limitations. Stud. Biophys. **110**: 43, 50.

Graetz, L. (1900). Ueber die Quincke'schen Rotationen im elektrischen Feld. Ann. Phys. **1**: 530, 541.

Hagedorn, R., and Fuhr, G. (1984). Calculation of rotation of biological objects in the electric rotation field. Stud. Biophys. **102**: 229, 238.

Hagedorn, R., Fuhr, G., Müller, T., and Gimsa, J. (1992). Traveling-wave dielectrophoresis of micro particles. Electrophoresis **13**: 49, 54.

Hertz, H. R. (1881). Über die Vertheilung der Electricität auf der Oberfläche bewegter Leiter. Wied. Ann. **13**: 266, 275.

Heydweiller, A. (1897). Ueber Rotationen im constanten electrischen Felde. Verh. Dtsch. Phys. Ges. **16**: 32, 36.

Heydweiller, A. (1899). 1. Ueber bewegte Körper im elektrischen Felde und über die elektrische Leitfähigkeit der atmosphärischen Luft. Ann. Phys. Chem. **69**: 531, 575.

Holzapfel, C. J., Vienken, J., and Zimmermann, U. (1982). Rotation of cells in an alternating electric field: Theory and experimental proof. J. Membrane Biol. **67**: 13, 26.

Hölzel, R. (1988). Sine-quadrature oscillator for cellular spin resonance up to 120 MHz. Med. Biol. Eng. Comp. **26**: 102, 105.

Hölzel, R., and Lamprecht, I. (1987). Cellular spin resonance of yeast in a frequency range up to 140 MHz. Z. Naturforsch. **42c**: 1367, 1369.

Huang, Y., Hölzel, R., Pethig, R., and Wang, X-B. (1992). Differences in the a.c. electrodynamics of viable and non-viable yeast cells determined through combined dielectrophoresis and electrorotation studies. J. Phys. Med. Biol. **37**: 1499, 1517.

Hub, H.-H., Ringsdorf, H., and Zimmermann, U. (1982). Rotation of polymerized vesicles in an alternating electric field. Angew. Chem. Int. Ed. Engl. **21**: 134, 135.

Irimajiri, A., Hanai, T., and Inouye, A. (1979). A dielectric theory of "multi-stratified shell" model with its application to a lymphoma cell. J. Theor. Biol. **78**: 251, 269.

Jones, T. B. (1984). Quincke rotation of spheres. IEEE Trans. Ind. Appl. IA-**20**: 845, 849.

Jones, T. B. (1986). Dielectrophoretic force in axisymmetric fields. J. Electrostat. **18**: 55, 62.

Kaler, K. V. I. S., and Johnston, R. H. (1985). Spinning response of yeast cells to rotating electric fields. J. Biol. Phys. **13**: 69, 73.

Krasny-Ergen, W. (1937). Der Feldverlauf im Bereich sehr kurzer Wellen, spontane Drehfelder. Hochfrequenztechn. Elektroakust. **49**: 195, 199.

Kumar, S., and Cho, D. (1990). A pertubation method for calculating the capacitance of electrostatic motors. IEEE **4**: 27, 33.

Küppers, G., Wendt, B., and Zimmermann, U. (1983). Rotation of cells and ion-exchange beads in the MHz-frequency range. Z. Naturforsch. **38c**: 505, 507.

Lampa, A. (1906). Ueber Rotationen im elektrostatischen Drehfelde. Wien. Ber. 2a, **115**: 1659, 1690.

Lang, V. V. (1906). Versuche im elektrostatistischen Drehfelde. Wien. Ber. **115**: 211, 222.

Lertes, P. (1921a). Untersuchungen über Rotationen von dielektrischen Flüssigkeiten im elektrostatischen Drehfeld. Z. Phys. **4**: 315, 336.

Lertes, P. (1921b). Der Dipolrotationseffekt bei dielektrischen Flüssigkeiten. Z. Phys. **6**: 56, 68.

Lovelace, R. V. E., Stout, D. G., and Steponkus, P. L. (1984). Protoplast rotation in a rotating electric field: The influence of cold acclimation. J. Membr. Biol. **82**: 157, 166.

Mehregany, M., Nagarkar, P., Senturia, St. D., and Lang, J. H. (1990). Operation of microfabricated harmonic and ordinary side-drive motors. IEEE **4**: 1, 8.

Melcher, J. R. (1974). Electric fields and moving media. IEEE Trans. Educ. E-17: 100, 110.

Melcher, J. R., and Taylor, G. I. (1969). Electrohydrodynamics: A review of the role of interfacial shear stresses. Annu. Rev. Fluid Mech. **1**: 111, 146.

Meyer, O. E. (1877). Die kinetische Energie der Gase. Anm. 3: 156. Breslau. *cit. after* Heydweiller, A. (1896). Ueber Rotationen im constanten electrischen Felde. Verh. Phys. Ges. Berlin **16**: 32, 36.

Mischel, M., and Lamprecht, I. (1980). Dielectrophoretic rotation in budding yeast cells. Z. Naturforsch. **35c**: 1111, 1113.

Mischel, M., and Lamprecht, I. (1983). rotation of cells in nonuniform alternating fields. J. Biol. Phys. **11**: 43, 33.

Mischel, M., and Pohl, H. A. (1983). Cellular spin resonance: Theory and experiment. J. Biol. Phys. **11**: 98, 102.

Mischel, M., Voss, A., and Pohl, H. A. (1982). Cellular spin resonance in rotating electric fields. J. Biol. Phys. **10**: 223, 226.

Müller, T., Fuhr, G., Geissler, F., and Hagedorn, R. (1988). Electrorotation measurements on mouse blastomers. Pages 179, 182. *In* Electromegnetic Fields and Biomembranes. Markov, M., and Blank, M., eds. Plenum Press, New York.

Müller, T., Fuhr, G., Hagedorn, R., and Göring, H. (1986). Influence of dielectric breakdown on electrorotation. Stud. Biophys. **113**: 203, 211.

Müller, T., Küchler, L., Fuhr, G., Schnelle, Th., and Sokirku, A. (1993). Dielektrische Einzelspektroskopie an Pollen verschiedener Waldbaumarten-Charakterisierung der Pollenvitalität. Silvae Genetica **42**: 311, 322.

Ogava, T. (1961). Measurement of the electrical conductivity and dielectric constant without contacting electrodes. J. Appl. Phys. **32**: 583, 592.

Okano, K. (1965). On the rotatory motion of dielectrics in static electric fields. Jpn. J. Appl. Phys. **4**: 292, 296.

Pastushenko, V. Ph., Kuzjmin, P. I., and Chizmadzhev, Yu. A. (1985): Dielectrophoresis and electrorotation: A unified theory of spherically symmetrical cells. Stud. Biophys. **110**: 51, 57.

Pethig, R. (1991a). Application of a.c. electrical fields to the manipulation and characterisation of cells. Pages 159, 185. *In* Automation in Biotechnology, Karube, I., ed. Elsevier, Amsterdam.

Pethig, R. (1991b). Biological electrostatics: Dielectrophoresis and electrorotation. Inst. Phys. Conf. Ser. No. **118**: 13, 26.

Pickard, W. F. (1961). On the Born-Lertes rotational effect. II Nuovo Cimento **21**: 316, 332.

Pickard, W. F. (1965). Electrical force effects in dielectric liquids. Progr. Dielec. **6**: 1, 39.

Pilwat, G., and Zimmermann, U. (1983). Rotation of a single cell in a discontinuous rotating electric field. Bioelectrochem. Bioenerg. **10**: 155, 162.

Pohl, H. A. (1978). Dielectrophoresis: The behavior of neutral matter in nonuniform electric fields. Cambridge University Press. Cambridge, London, New York, Melbourne.

Pohl, H. (1983a). The spinning of suspended particles in a two pulsed three-electrode system. J. Biol. Phys. **11**: 66, 68.

Pohl, H. (1983b). Cellular spin resonance: A new method for determining the dielectric properties of living cells. Int. J. Quantum Chem. **10**: 161.

Pohl, H. (1986). Cellular spin resonance spectrometer. US Patent Nr. 4, 569, 741. (11.2. 1986).

Pohl, H., and Crane, J. S. (1971). Dielectrophoresis of cells. Biophys. J. **11**: 711, 727.

Quincke, G. (1896). Ueber Rotationen im constanten elektrischen Felde. Wied. Ann. **59**: 417, 486.

Richardson, S. W. (1927). Rotation of dielectric bodies in electrostatic fields. Nature **119**: 238.

Sauer, F. A. (1983). Forces on suspended particles in the electromagnetic field. Pages 134, 143. In Coherent Excitation in Biological Systems. Fröhlich, H., and Kremer, F., eds. Springer-Verlag, Berlin-Heidelberg.

Sauer, F. A., and Schlögl, R. W. (1985). Torques exerted on cylinders and spheres by external electromagnetic fields. A contribution to the theory of induced cell rotation. Pages 203, 251. In Interactions Between Electromagnetic Fields and Cells. Chiabrera, A., Nicolini, C., and Schwan, H. P., eds. Plenum, New York.

Schnelle, Th., Hagedorn, R., Fuhr, G., Fiedler, S., and Müller, T. (1993). Three-dimensional electric field traps for manipulation of cells: Calculation and experimental verification. Biochim. Biophys. Acta **1157**: 127, 140.

Schwan, H. (1957). Electrical properties of tissue and cell suspensions. Pages 147, 209. In Advances in Biological and Medical Physics. Lawrence, J. H., and Tobias, C. A., eds. Academy Press, New York.

Schwan, H. P. (1987). Electro-rotation: The context with respect to other pondero-motive effects. Pages 511, 513. In Proceedings of the 13th Annual NE Bioengineering Conference 1990. Forster, K. R., ed. IEEE, New York.

Schweidler, E. v. (1897). Über Rotationen im homogenen elektrischen Felde. Wien. Ber. **106**: 526, 532.

Schweidler, E. v. (1907). Studien über die Anomalien im Verhalten der Dielektrika. Wien. Ber. **116**: 1019, 1080.

Secker, P. E., and Belmont, M. R. (1970). A miniature multipole liquid-immersed dielectric motor. J. Phys. D: Appl. Phys. **3**: 216, 220.

Secker, P. E., and Scialom, I. N. (1968). A simple liquid-immersed dielectric motor. J. App. Phys. **39**: 2957, 2961.

Simpson, P., and Taylor, R. J. (1971). Characteristic rotor speed variations of a dielectric motor with a low-conductivity liquid. J. Phys. D: Appl. Phys. **4**: 1893, 1897.

Sokirku, A. V. (1992). Electrorotation of axial symmetric cell. Biol. Mem. **6**: 587, 600.

Steinmetz, C. P. (1892). Dielektrische Hysteresis, der Energieverlust in dielek-trischen Medien unter dem Einfluss eines wechselnden elektrostatischen Feldes. Elektrotechn. Z. **13**: 227, 228.

Stratton, J. A. (1941). Electromagnetic Theory. McGraw-Hill, New York.

Sumoto, I. (1956). Rotary motion of dielectrics in a static electric field (in Japanese, English summary). Rep. Sci. Res. Inst. **32**: 41, 56.

Tai, Y., Fan, L., and Muller, R. (1989). IC-processed micro-motors: Design, technology and testing. Pages 1, 6. In Proceedings IEEE Micro Electro Mechanical Systems Workshop, Salt Lake City.

Teixera-Pinto, A. A., Nejelski, L. L., Cutler, J. L., and Heller, J. H. (1960). The behaviour of unicellular organism in an electromagnetic field. Exp. Cell Res. **20**: 548, 564.

Trimmer, W. S. N., and Gabriel, K. J. (1987). Design considerations for a practical electrostatic micro-motor. Sensors and Actuators **11**: 189, 206.

Turcu, I., and Lucaciu, C. M. (1989). Electrorotation: A spherical shell model. J. Phys. A: Math. Gen. **22**: 995, 1003.

Vedy, L. G. (1931). On the rotation of dielectrics in electrostatic fields and related phenomena. Proc. Cambridge Phil. Soc. **27**: 91, 102.

Wang, X.-B., Huang, Y., Gascoyne, P. R. C., Becker, F. F., Hölzel, R., and Pethig, R. (1994). Changes in friend murine erythroleukaemia cell membranes during induced differentiation determined by electrorotation. Biochim. Biophys. Acta **1193**: 330, 344.

Washizu, M. (1989). Electrostatic manipulation of biological objects in microfabricated structures. Third Toyota-Conference, 1, 19.

Washizu, M., Kurahashi, Y., Iochi, H., Kurosawa, O., Aizawa, S., Kudo, S., Magariyama, Y., and Hotani, H. (1993). Dielectrophoretic measurement of bacterial motor characteristics. IEEE Trans. Ind. Appl. **29**: 286, 294.

Wicher, D., and Gündel, J. (1985). Berechnungen zur Rotation biologischer Einzelzellen im rotierenden elektrischen Feld unter Zugrundelegung eines ein-bzw. zweischaligen Modells. Forschungsergenbnisse Friedrich-Schiller-Universität Jena **26**: 1, 16.

Wicher, D., Gündel, J., and Matthies, H. (1986). Measuring chamber with extended applications of the electro-rotation. Alpha and beta dispersion of liposomes. Stud. Biophys. **115**: 51, 58.

Ziervogel, H., Glaser, R., Schadow, D., Heymann, S. (1986). Electrorotation of lymphocytes: The influence of membrane events and nucleus. Biosci. Rep. **6**: 973, 982.

Zimmermann, U. (1982). Electric field-mediated fusion and related electrical phenomena. Biochim. Biophys. Acta **694**: 227, 277.

Zimmermann, U., and Arnold, W. M. (1983. The interpretation and use of the rotation of biological cells. Pages 211, 221. *In* Coherent Excitation in Biological Systems. Fröhlich, H., and Kremer, F., eds. Springer-Verlag, Berlin.

Zimmermann, U., Vienken, J., and Pilwat, G. (1981). Rotation of cells in an alternating electric field: The occurrence of a resonance frequency. Z. Naturforsch. **36c**: 173, 177.

CHAPTER 4

Dielectrophoresis: Effect of Nonuniform Electrical Fields on Cell Movement

F. J. Iglesias

C. Santamaría

F. J. Asencor

A. Domínguez

ABSTRACT

Dielectrophoresis, or translational movement of neutral matter due to the action of a nonuniform electrical field, has been applied intensively to the study of biological particles and colloidal dispersions. Dielectrophoresis can provide details of the cell or particle surface charge and dielectrical properties. The dielectrophoretic technique may also be used in practical applications, such as particle separation and levitation, as well as the dielectrical characterization of irregularly shaped solids. This chapter focuses on the biotechnological applications of dielectrophoresis such as cell separation and cell characterization. Currently, the two most interesting aspects of dielectrophoresis are its applications to electrofusion and to the analysis of cell orientation. Both these phenomena have been analyzed using microorganisms (particularly yeasts), because of the reproducibility of their growth characteristics and the homogeneity of their cell populations.

INTRODUCTION

The motion of an uncharged particle that results from its interaction with a nonuniform electrical field is termed dielectrophoresis (Pohl, 1951, 1958). The origin of this force lies in the interaction of the field with the dipoles induced by the field on the particle. In principle, the motion of the particle is independent of the direction of the field; a freely suspended particle will move in the direction in which the field increases. The dielectrophoretic technique can be used with inanimate systems. For example, it is possible to filter nonconductive liquids (Fielding et al., 1975), to separate mineral powder mixtures into their component parts (Verschure and Ijlst, 1966), and to gain insight into the behavior of disperse systems or colloids. This can be a powerful tool in the interpretation of the anomalous polarization that occurs in heterogeneous systems (Santamaría et al., 1985). Dielectrophoresis has recently gained interest among cell biologists, biophysicists, and biotechnologists for several reasons. Using dielectrophoresis, it is possible to analyze the frequency spectra of bacteria and yeasts (Pohl and Crane, 1971), the behavior of living and nonliving cells (Pohl, 1978; Iglesias et al., 1984), together with the response of cellular organelles (Ting et al., 1971). Dielectrophoresis can reflect the electrical properties of any single cell and, for example, Kaler and Jones (1990) have measured the dielectrophoretic levitation spectra of *Canola* protoplasts and ligament fibroblast cells. These authors have shown that it is possible to measure the effective membrane capacitance of single cells, and the values obtained are in agreement with those obtained with other existing electrical methods. The study of the mechanisms by which electrical fields induce movement is also of considerable theoretical interest.

One of the most recent applications of dielectrophoresis has been to aggregate cells of the same or of different types to obtain cellular fusions. If cells or membranes are first aligned into close contact by dielectrophoresis, then an electrical signal may induce fusion (Sowers, 1989; Teissié et al., 1989). In our laboratory, attention has been focused on the study of the dielectrophoretic behavior of microorganisms as models in order to understand their application in biotechnological processes.

CELL ORIENTATION

The techniques of cell fusion have acquired considerable importance in recent years, not only because they are an excellent tool for understanding and interpreting fundamental questions in cell biology, such as gene expression, but also because of their practical use in biotechnology. Fusions that use chemical agents (e.g., polyethylene-glycol) can be improved by the use of electrofusion and electroporation. Among the problems inherent to these techniques, however, is that of situating the cells to be electrofused in a preset orientation. Thus, study of cell orientation in electrical fields is of considerable interest, regardless of the type of cell symmetry.

Cell Orientation of Nonspherical Cells

As is well known, a nonspherical particle immersed in a medium with different dielectrical properties tends to become oriented with the electrical field. Due to the lack of spherical symmetry, the potential energy of the system will depend on the relative position of the particle with respect to the lines of the electrical field. The stabilized equilibrium position will depend on the orientation for which the potential energy is minimal. If the particle were ellipsoid in shape and the medium were a perfect dielectrical medium, this position would correspond to the one where the major axis of the particle would be aligned with the lines of the electrical field. In reality, however, both the medium and the particle should be considered as dielectrical, but with losses; that is, it is necessary to take into account the electrical conductivity of the medium, such that in the formulation of the problem complex magnitudes will arise; in particular, the dielectrical constant will change from being a real magnitude (ε) to a complex number ($\tilde{\varepsilon}$),

$$\tilde{\varepsilon} = \varepsilon - j\frac{\sigma}{\omega}$$

where σ is the electrical conductivity, ω is the frequency of the field, and $j = \sqrt{-1}$. The dependence on energy is a complex function of the frequency of the alternating field. In such cases, the equilibrium position does not necessarily have to coincide with the major symmetry axis parallel to the direction of the field but rather will correspond to intermediate positions. Different authors (Teixeira-Pinto et al., 1960; Griffin and Stowell, 1966; Friend et al., 1975; Vienken et al., 1984; Iglesias et al., 1985; Mishima and Morimoto, 1989) have demonstrated this type of orientation in different microorganisms. Our team (Iglesias et al., 1985, 1989) have shown that in the 1–20 MHz frequency range and with an electrical conductivity of the medium between 1 mS/m and 25 mS/m, living cells always become oriented with their longest axis parallel to the field lines. In contrast, in suspensions of dead cells, the orientation of the major axis may lie either parallel or perpendicular to the direction of the field. In fact, according to the frequency of the field applied and the conductivity of the medium, both orientations may coexist. In our work, the systems studied have involved three bacteria (*Escherichia coli*, *Bacillus subtilis*, and *Bacillus megaterium*), a yeast (*Schizosaccharomyces pombe*), and a spore of the fungus *Phycomyces blakesleeanus*. The microorganisms were killed by different methods, including autoclaving at 121°C, permeabilization with toluene–ethanol, ultraviolet (UV) light and chemical reagents. The results, common to the different treatments, suggest that the dielectrophoretic transition (parallel–perpendicular) is due to the fact that an exchange of ions occurs between the cell cytoplasm and the external medium. This can be confirmed, because in all cases, the membrane is seen to have become permeabilized.

Schwarz et al. (1965) and Saito et al. (1966) conducted theoretical studies on the stability of a given orientation for ellipsoid particles with and without shells immersed in a medium with different dielectrical properties. The direction of stable orientation is determined by the material involved and also by the frequency of the fields. The change, in the stable direction on modifying the frequency and the conductivity of the medium, may occur, in principle, either as a gradual change or as a sudden jump of 90°. Saito et al. in 1966 analyzed the stable orientation for ellipsoid particles with and without shells by the minimum energy principle. According to their theory, turnover can only occur if ellipsoid particles with a shell of symmetrical composition and parameters of biological interest are considered. For field intensities that are not sufficiently high, however, this sudden jump may be blurred by geometric parameters, dispersion, friction on the microscope slide, or thermal agitation.

Assuming that microorganisms studied in our laboratory can be considered as revolution ellipsoids with a shell representing the cell membrane, the theory of Saito et al. (1966) interprets our results well in qualitative terms. Starting from the expression for the potential energy stored in the system given by Schwarz et al. (1965), Saito et al. (1966) obtained a value for the potential energy of each orientation of a revolution ellipsoid (Fig. 4-1) given by

$$W_i = \tfrac{1}{2}E^2\lambda_i \qquad (i = a, b)$$

Subindex *a* refers to the case in which the direction of the principal axis coincides with that of the electrical field, and subindex *b* refers to the case when it is perpendicular to the electrical field. *E* is the strength of the electrical field, and λ_i depends on the geometry of the particle and on the

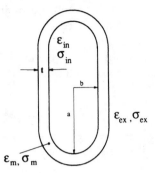

Figure 4-1. One-shell electric model of a revolution ellipsoid. The dielectric permittivity ε, and electric conductivity σ, in different phases are specified. The subindices in, m, and ex, denote, respectively, cell interior, cell membrane and suspending medium; a and b are major and minor axes of the ellipsoid and t is the membrane thickness. Source: Biophys. J. *64*: 1626–1631, with permission of the Biophysical Society.

electrical parameters through

$$\lambda_i = \frac{4}{3}\pi ab^2\varepsilon_{ex}\left[\frac{\varepsilon_{ex} - \varepsilon_i}{\varepsilon_{ex} - (\varepsilon_{ex} - \varepsilon_i)L_i}\right.$$

$$\left. + \frac{B_i}{[\sigma_{ex} - (\sigma_{ex} - \sigma_i)L_i]^2 + [\varepsilon_{ex} - (\varepsilon_{ex} - \varepsilon_i)L_i]^2\omega^2}\right]$$

$$B_i = \frac{2\sigma_{ex}\varepsilon_{ex}(1 - L_i) + (\sigma_{ex}\varepsilon_i + \sigma_i\varepsilon_{ex})L_i}{\varepsilon_{ex}^2 - (\varepsilon_{ex} - \varepsilon_i)\varepsilon_{ex}L_i}(\sigma_{ex}\varepsilon_i - \sigma_i\varepsilon_{ex})$$

where *a* and *b* are the major and minor semiaxes, respectively, of the revolution ellipsoid. ε_i and σ_i (permittivity and electrical conductivity) are given by the real and imaginary part of the following.

$$\varepsilon_i - j\frac{\sigma_i}{\omega} = \frac{(2 - e_iM_i)\left(\varepsilon_{in} - j\frac{\sigma_{in}}{\omega}\right) + e_iM_i\left(\varepsilon_m - j\frac{\sigma_m}{\omega}\right)}{e_iN_i\left(\varepsilon_{in} - j\frac{\sigma_{in}}{\omega}\right) + (2 - e_iN_i)\left(\varepsilon_m - j\frac{\sigma_m}{\omega}\right)}\left(\varepsilon_m - j\frac{\sigma_m}{\omega}\right)$$

$$M_a = \frac{2}{b^2}; \qquad M_b = \frac{1}{a^2} + \frac{1}{b^2}; \qquad N_a = \frac{1}{a^2}; \qquad N_b = \frac{1}{b^2};$$

$$e_a = 2at; \qquad e_b = 2bt$$

t is the thickness of the membrane and the subindexes (ex) and (in) refer to the exterior and interior media of the cell, respectively; subindex m refers to the cell membrane.

Once solved, the parameters L_i given by Saito et al., 1966 take the following values for our revolution ellipsoid:

$$L_a = \frac{b^2}{a^2 - b^2} \left[\frac{a}{2\sqrt{a^2 - b^2}} \ln \left(\frac{a + \sqrt{a^2 - b^2}}{a - \sqrt{a^2 - b^2}} \right) - 1 \right]$$

$$L_b = \frac{ab}{a^2 - b^2} \left[\frac{a}{2b} - \frac{b}{4\sqrt{a^2 - b^2}} \ln \left(\frac{a + \sqrt{a^2 - b^2}}{a - \sqrt{a^2 - b^2}} \right) \right]$$

The numerical values used in the calculation program compiled by us were as follows: geometric parameters (semiaxes of the ellipsoid) $a = 5$ μm; $b = 1.13$ μm, membrane thickness $t = 80$ nm. These parameters correspond to *S. pombe* and were obtained by electron microscopy. The electrical parameters used were those customarily used in cellular physiology (Saito et al., 1966; Pohl, 1978). Dielectrical permittivity inside the cell was 60 ε_0, that of the membrane 11 ε_0, and that corresponding to the suspending medium, which was an aqueous salt solution, 78 ε_0, being $\varepsilon_0 = 8.85 \times 10^{-12}$ F/m, the permittivity of the free space. The electrical conductivity of the medium was controlled externally over the 1–25 mS/m range. Additionally, the electrical conductivities of the membrane and of the cytoplasm are necessary for computing the energy of each orientation. Unfortunately, there are no experimental results for these parameters. In the case of the membrane, a value of 5×10^{-4} S/m was taken, which is of the same order as that used by Pohl (1978) for yeast and by Hatakeyama et al. (1987) for *Paramecium*. The value used by our team is slightly higher than that used by these authors because we used dead microorganisms. It therefore seems probable that the pores of the membrane may provide an increase in ionic bonds, making them more conductive.

We have calculated the distribution of orientations as a function of the frequency of the electric field and of the conductivity of the external medium for a conductivity value in the cell cytoplasm of 4×10^{-2} S/m. The results are shown in Figures 4-2 and 4-3. The quantitative agreement with our experimental results for *S. pombe* is noteworthy (see Fig. 9 in Iglesias et al., 1989). Two clearly differentiated regions can be observed, one of them closest to the ordinate axis, where the energy corresponding

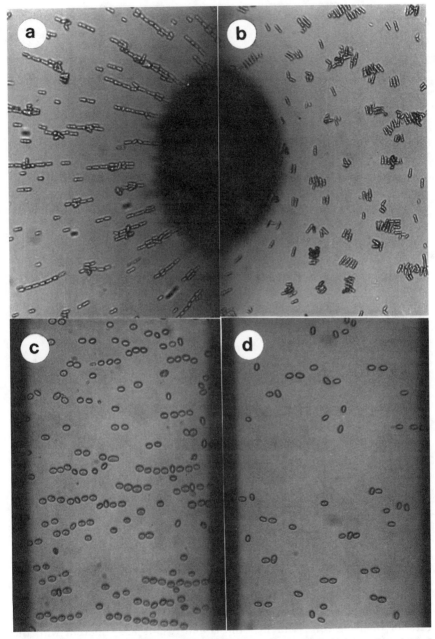

Figure 4-2. Sequential photomicrographs of cellular orientation as a function of the frequency of the AC field. (a, b) non-living *Schizosaccharomyces pombe* cells under a non-uniform electric field. (c, d) *Phycomyces blakesleeanus* spores under a uniform electric field. Frequency (a and c, parallel), 1 MHz; (d, mixed), 15 MHz; (b, perpendicular), 30 MHz. Electrical conductivity, 2×10^{-3} Sm^{-1}.

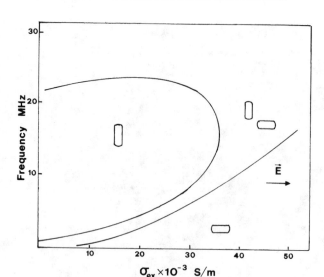

Figure 4-3. Computed orientation of non-living cells of *Schizosaccharomyces pombe* in an alternating electric field as a function of frequency and conductivity. The model has been elaborated as described in the text with our data (Iglesias et al., 1985).

to the perpendicular orientation is noticeably lower than the parallel orientation energy, and another region close to the abscissa axis, where the opposite is the case. The stable equilibrium position is that corresponding to the parallel orientation. There is also an intermediate region in which the energies of the two orientations are similar (neither of them predominating) such that they are able to coexist. This "mixing" region corresponds to the situation in which the difference in energies for both orientations is comparable to the thermal agitation energy $1/2\ kT$, where k is the Boltzmann constant and T is the absolute temperature corresponding to the rotation of the particle.

It should be noted that the qualitative agreement between the theoretical and experimental values can only be achieved if a value of 4×10^{-2} S/m is chosen for the electrical conductivity inside the particle. This conductivity is similar, but slightly higher, than that of the external medium and, in any case, much lower than that usually used for the interior of living cells. Our hypothesis and results are coherent with the notion that permeabilization of the plasma membrane permits a passive flow of ions and metabolites that tends to equalize the conductivity of both media (as an example, the penetration of methylene blue by dead cells). Such alterations in the membrane and some of those occurring in the cell wall (e.g., a partial solubilization of galactomananan) allow cells of *S. pombe* to keep their shape. Again, this result is coherent with the hypothesis advanced by Pohl (1978) to the effect that in killed cells the electrical conductivity of the cytoplasm is equal to that of the external medium. The orientation values obtained from *S. pombe* (Fig. 4-2, and Fig. 4-3), either experimentally or with our computer program, take the same aspect as those described by Hatakeyama et al. (1987; see Fig. 4-8) for *Paramecium*. Two main differences should be pointed out, however. The first is that with our experimental devices, we were unable to obtain a perpendicular orientation to the field lines with living *S. pombe* cells. The second is that with dead cells, we found values for mixed orientation (coexistence of cells, parallel or perpendicular) that could also be predicted in our computer program. Similar results have not, however, been described for *Paramecium*, either in the experimental results or in the theoretical program described by Hatakeyama et al. (1987). The fact that reorientation of living cells is not observed in the frequency range studied can readily be attributed to the high conductivity of the cytoplasm. This conductivity is far in excess of that of the external medium, such that frequencies of the alternating field much higher than those that can be achieved with our experimental device would be necessary to observe the reorientation phenomenon in live cells of *S. pombe* or of other microorganisms.

Cell Orientation of Spherical Cells

The foregoing theoretical model is evidently unable to account for the phenomena of cell orientation when the particle has spherical symmetry (i.e., yeasts or plant protoplasts). Our hypothesis is that the orientation of spherical particles constituting pearl chains is random. Hence, for the purposes of performing a controlled electrofusion, some kind of marker or signal should be incorporated into a plasma membrane in order to break the spherical symmetry. An example of how yeast (*Saccharomyces cerevisiae*) can be manipulated from the genetic viewpoint to incorporate and hence to monitor small asymmetries has been described by Hartwell et al. (Cross et al., 1988; Jackson and Hartwell, 1990a, 1990b; Jackson et al., 1991). These authors have shown that yeast cells of opposite sexual type,

Mat a and Mat α, are capable of choosing a mating partner from a group of neighboring cells, depending on a pheromone gradient in their environment. Defects in actin, myosin 2, and clathrin heavy chains are involved in mating partner discrimination. Thus, mating partner discrimination requires the reorganization of the cytoskeleton of the cell, such that spherical or quasispherical cells become oriented, permitting a series of events leading to a specific "conjugation-type selection" that could be used for achieving specific electrofusion of spherical cells. Another type of manipulation for achieving the oriented fusion of S. pombe protoplasts by digestion of the apical zone of the cell wall has been reported by Vondrejs et al. (1990).

Uniform and nonuniform alternating electrical fields cause another type of movement different from that observed in dielectrophoresis; namely, electrorotation of cells around one of their symmetry axes. This type of movement is also seen in rotating electrical fields (Fuhr et al., 1986; Mischel and Lamprecht, 1983). The rotation speed is highly susceptible to the frequency of the field applied. In some cases, the field may even invert the direction of rotation, from counterclockwise to clockwise (Pollock et al., 1987). Electrorotation can be due to dipole interactions, as happens in dielectrophoresis. When the frequency of the alternating field is modified, a resonance frequency, at which all the cells in suspension rotate around a normal axis to the lines of the field, is observed (Zimmermann et al., 1981). This resonance frequency strongly depends on the type of microorganism under study. Electrorotation can also be altered by the addition of certain special substances such as phenols, as indicated by Arnold et al. (1988a, 1988b). These authors used electrorotation to detect membrane differences among members of the same cell population. Rotation phenomena are strongly dependent upon the difference in the complex dielectrical constants between the inside and the outside of the cell. Slight modifications or changes in the cell membrane, for example, changes in permeability because of damage to the membrane, mean that electrorotation may be altered or that it may disappear altogether (Glaser et al., 1983; Hatakeyama and Yagi, 1990). Thus, analysis of this type of cellular movement, electrorotation, together with dielectrophoretic measurement, constitute a useful tool for obtaining experimental results concerning the electric conductivity of membranes in vivo (Holzapfel et al., 1982; Zimmermann and Arnold, 1983; Marszalek et al., 1991; Gimsa et al., 1991) and may assist in understanding problems inherent to cell electrofusion. It should be emphasized however, that, from the point of view of cell electrofusion, this type of movement should be avoided, since rotation of the cells in pearl chains interrupts contact among the membranes of adjacent cells.

FACTORS AFFECTING CELL MOVEMENT

Theoretical Considerations

When a neutral particle is subjected to the effect of a nonuniform electrical field, a force is exerted on that particle. Because of the polarizability of the particle, it can be considered as a dipole with dipole moment, **p**. The nonuniform nature of the field means that there will be a net force $\mathbf{F}_d$ that displaces the particle. The value of this force is

$$\mathbf{F}_d = (\mathbf{p} \cdot \nabla)\mathbf{E}$$

where $\mathbf{E}$ represents the intensity of the applied external field. Accepting linearity between **p** and **E**, this expression is transformed into

$$\mathbf{F}_d = \tfrac{3}{2}\upsilon X \nabla |\mathbf{E}|^2 \tag{1}$$

where υ is the volume of the particle and X will be called the effective polarizability of the particle. For the case in question of alternating electrical fields, X is a complex function of the electrical properties of the medium and of the particle. These electrical properties are complex permittivities of both materials:

$$\tilde{\varepsilon} = \varepsilon - j\frac{\sigma}{\omega}$$

where ε is the dielectrical permittivity, σ is conductivity, and ω is the angular frequency of the applied electrical field. For a homogeneous particle of radius R and with a permittivity ε_1 suspended in a homogeneous medium of permittivity ε_2, authors have offered different expressions for the real value of X. Thus, Jones and Kallio (1979) and Benguigui and Lin (1982) have proposed

$$X = \mathrm{Re}\left\{\tilde{\varepsilon}_1 \frac{\tilde{\varepsilon}_2 - \tilde{\varepsilon}_1}{\tilde{\varepsilon}_2 + 2\tilde{\varepsilon}_1}\right\}$$

Pohl (1978) prefers the expression

$$X = \text{Re} \left\{ \tilde{\varepsilon}_1^* \frac{\tilde{\varepsilon}_2 - \tilde{\varepsilon}_1}{\tilde{\varepsilon}_2 + 2\tilde{\varepsilon}_1} \right\}$$

where Re indicates the real part and $\tilde{\varepsilon}^*$ is the conjugated complex of $\tilde{\varepsilon}$.

This disparity arises from different theoretical treatments of the problem and, in any case, the experimental results obtained to date do not help to clarify the controversy.

In the light of these considerations, our attention will focus on the value of X, without referring to its relationship with the electrical properties. From Equation 1, the dielectrical force depends directly on the gradient of the square of the intensity of the electrical field. The presence of the square makes the dielectrophoretic force independent of the sign of the electrical field, it being possible to use, as in our case, alternating fields. When this force is maximized, the appropriate electrode symmetry should be that which offers the best nonhomogeneity or divergence from the electrical field. When this force is positive ($X > 0$), it causes a displacement of the particles in suspension toward the zones where the field is most intense, the particles being deposited on the electrodes to form pearl chains. The configuration of the electrodes that generates field inhomogeneity to avoid errors in measurement should be such that the deposition of pearls chains adhering to the electrodes will be as homogeneous and symmetric as possible. The distribution used in our work was thus the one that would generate spherical symmetry (Fig. 4-4d). For this type of symmetry, it is easy to obtain the value $\nabla |\mathbf{E}|^2$, with the following value for force:

$$\mathbf{F}_d = 4\pi R^3 X \frac{r_1^2 r_2^2 V^2}{r^5 (r_2 - r_1)^2}$$

where V is the applied root mean square (rms) voltage and r_1 and r_2 are the internal and external radii of the spherical electrodes. This dielectrical force of attraction to the electrodes is not the only one to which the particle is subject, since it is also necessary to take into account the drag force, which in our experimental device is given by Stokes law:

$$\mathbf{F}_v = -6\pi\eta R\mathbf{v}$$

where η is the viscosity of the fluid and v is the relative velocity of the particle. The motion equation is

$$\mathbf{F}_d + \mathbf{F}_v = m\ddot{r}$$

where $\mathbf{F}_d$ represents the dielectric force, $\mathbf{F}_v$ the drag force, and $\ddot{r}$ the acceleration of the particle. Having solved this differential equation, the trajectory of the particle and the length of the pearl chains adhering to the electrode can be calculated. Study of this dielectrophoretic deposition when different biological and physical parameters are modified allows us to interpret and analyze the mechanism of polarization occurring in the membrane and double ionic layer. The length of the chains, known as yield Y, was the object of our experimental measurement. The parameters on which this depends are analyzed below.

EXPERIMENTAL DEVICES

One of the parameters most difficult to measure is the yield Y or dielectrophoretic collection rate, defined as the average length of the chains of cells collected after a fixed time interval (Pohl, 1978). Several types of electrode configurations have been used (Fig. 4-4). Many of our results have been obtained with a pin–pin type electrode arrangement (Fig. 4-4c; Fig. 4-11). The platinum wire diameter in this was 65 μm and the electrodes were 1 mm apart. The pin types were rounded (approximately spherical), and the electrodes were mounted directly above the surface of the slide.

Another electrode configuration that affords even better results, especially with dead cells, has been described (Asencor et al., 1990). Dielectrophoresis was carried out using two electrodes in a device consisting of two perfectly fitting pieces. The lower part supported a silver electrode, with a central orifice of 5 mm diameter acting as a chamber. A platinum electrode, with a spherical end 230 μm in diameter, was suspended from the upper part into the chamber. The dielectrophoresis assembly is shown in Fig. 4-5a. Fig. 4-5b and Fig. 4-6 give details of the chamber. The ratio between the diameter of the external (like a sphere of infinite radius) and the internal electrode, meaning that spherical symmetry predominates over the cylindrical symmetry in the proximity of the inner electrode, is the principal difference with respect to the cylindrical chamber used previously (Dimitrov et al. 1984; Dimitrov and Zhelev, 1987). When this spherical symmetry is achieved, the aspect of yield is regular, permitting accurate measurement of the pearl chain length (Fig. 4-4d).

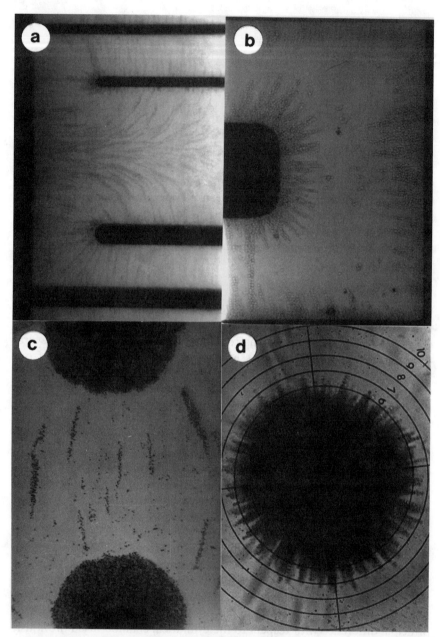

Figure 4-4. Photomicrographs of yeast collected on different experimental devices. a) and b) meandering chamber; c) dielectrophoretic deposit after disconnecting the electric field, (pin–pin type); d) dielectrophoretic deposit in our experimental device (see text and Asencor et al., 1990). Source: Biophys. J. *64*: 1626–1631, with permission of the Biophysical Society.

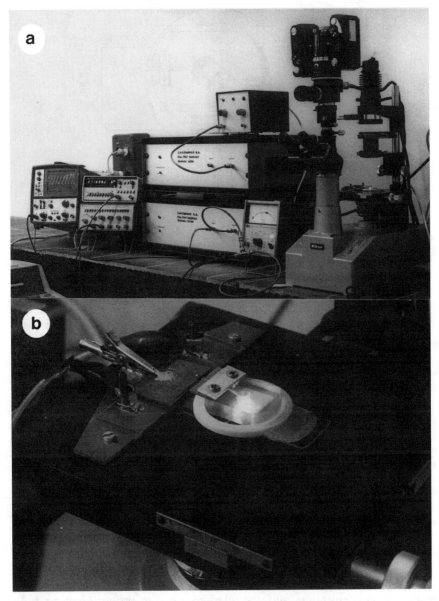

Figure 4-5. a) Dielectrophoretic assembly; b) detail of chamber.

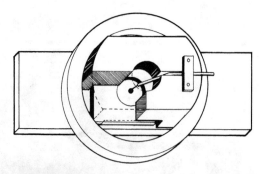

Figure 4-6. Basic design of the spherical dielectrophoretic chamber.

The orientation of microorganisms and spores in a uniform electrical field was studied in an assembly consisting of a microscope slide onto which two electrodes 0.75 mm apart had been painted with silver, as described by Teixeira-Pinto et al. (1960) (See Fig. 4-2). A small drop (50 μL) of the suspension to be examined was placed between the tips of the electrodes. Bacteria, yeasts, and fungi, power supplies, culture media, and the growth conditions of cells and spores, together with killing methods. UV irradiation and toxic treatments have been described previously in detail (Domínguez et al., 1980; Iglesias et al., 1984, 1985, 1989; López et al., 1984, 1985; Asencor et al., 1990, 1993).

Cell Concentration and Time of Collection

Crane and Pohl (1972) predicted that the yield of collected particles or cells is linear with respect to cell concentration and with respect to the voltage rms applied. Also, Y is proportional to the square root of the elapsed time and inversely proportional to the square root of the viscosity of the medium. Although the validity of such an approach has been demonstrated by Pohl (1978) for spherical cells, yeasts (live or dead), and protoplasts, many cells in nature have other shapes, being ellipsoid [such as *Yarrowia lipolytica* (Fig. 4-7b) or spores of *P. blakesleeanus* spores] or rodshaped [such as *E. coli*, *B. megaterium* or *S. pombe* (Fig. 4-7c)], and have different sizes. Our results show that the variations occurring in Y_R with cell concentration and the elapsed time is proportional to both parameters and is independent of particle size and shape (Iglesias et al., 1984), in agreement with the prediction of Pohl (1978). We also obtained a linear correspondence between the average length of the chains and the rms voltage, irrespective of cell size or the geometrical form of the

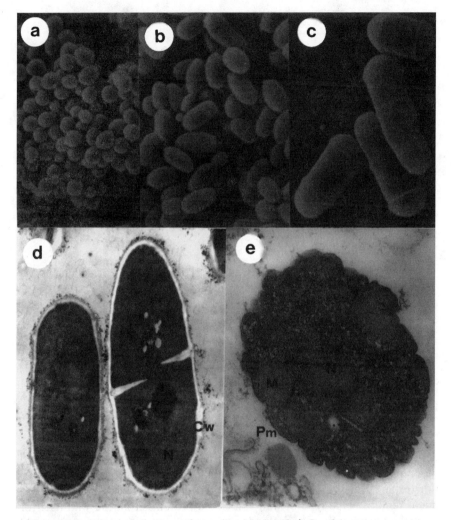

Figure 4-7. Scanning electron photomicrographs of a) *Saccharomyces cerevisiae*; b) *Yarrowia lipolytica* and c) *Schizosaccharomyces pombe* and transmission electron micrographs of d) *S. pombe* and e) *Y. lipolytica* protoplasts. Cw = cell wall; Pm = plasma membrane; N = nucleus, M = mitochondria.

microorganism (Fig. 4-8; Iglesias et al., 1984). Thus, cell concentration, the collection time, and rms voltage must be determined for each given microorganism and for each experimental set up.

A point that should be stressed is that the voltage usually used in our experiments does not damage either cells or protoplasts. This is shown by

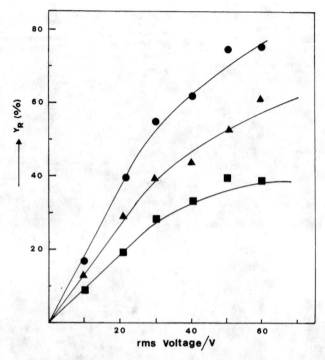

Figure 4-8. Dependence of yield on voltage for *Schizosaccharomyces pombe* (■) c = 7.2 × 10⁶; (▲) c = 11 × 10⁶; (●) c = 14.5 × 10⁶ cells/ml. Frequency was 1 MHz and electrical conductivity 2×10^{-3} Sm^{-1}. *Source: Biochim. Biophys. Acta*, 1984, **804**: 221–229, with permission of Elsevier Science Publishers.

determining the number of colonies before and after the pulse and observing that it remains constant. Similar results have been reported by Förster and Emeis (1985). Despite this, the possibility exists that the pulse might remove some type of wall-associated proteins from the wall itself or from the periplasmic. The influence of viscosity on dielectrophoretic yield has also been measured. Cells were suspended in several different media instead of in water and at different sugar concentrations. A decrease in yield was found when the viscosity of the medium was increased, in agreement with the hypothesis advanced by Pohl (1978).

Frequency and Conductivity

The dependence of yield on field frequency (yield spectrum) is probably the most important parameter for obtaining data concerning the behavior of different kinds of microorganisms. We have shown that by maintaining voltage and collection time and by fixing the conductivity, a similar dependence of yield on frequency (the same spectrum) is obtained for all of the several classes of microorganisms assayed. The yield spectrum showed a maximum between 0.1 and 1 MHz. This maximum was independent of the kind of microorganism studied, whether prokaryotic, such as *E. coli* or *Citrobacter freundii*, or eukaryotic such as yeast (*S. cerevisiae, S. pombe, Kloeckera apiculata*, or *Y. lipolytica*), fungal spores (*Verticillium chlamidosporium* or *P. blakesleeanus*), protoplasts (*S. cerevisiae* or *Y. lipolytica*), and human lymphocytes (Fig. 4-9). A slight displacement of the maximum toward lower frequencies was obtained for smaller sizes.

From our findings, it could be deduced that the dielectrophoretic behavior obtained is of the Maxwell–Wagner type (interfacial polarization). The only experimental discrepancy was the decrease in yield observed when cations of different valencies were added to the cell suspensions,

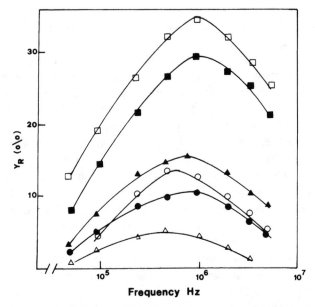

Figure 4-9. Frequency dependence of dielectrophoretic yield (Y_R). Electrical conductivity, 2×10^{-3} Sm^{-1}; V = 30 V rms, except for *Phycomyes blakesleeanus* and *Schizosaccaromyces pombe* = 20 V rms. Collection time, 2 min. (□) *P. Blakesleeanus* 1.2×10^{7} spores/ml; (■) *S. pombe*, 1.5×10^{7} cells/ml; (▲) *Saccharomyces cerevisiae* 1×10^{7} cells/ml; (●) and 1×10^{7} protoplasts/ml; (○) *Escherichia coli*, 1×10^{9} cells/ml; △, *S. cerevisiae*, 1×10^{7} dead cells/ml (by autoclaving).

maintaining a constant value of electrical conductivity (López et al., 1984). Also, a small displacement of the maximum and a slight increase in yield were seen when electrical conductivity was decreased (Iglesias et al., 1989).

Cell Type and Physiological Status

In our experimental work we used different types of cells that exhibit fundamental differences from the point of view of structure and metabolic state. We have worked with different types of prokaryotic microorganisms such as *E. coli* or *B. magaterium*, whose main difference with respect to eukaryotic organisms lies in the absence of an interior compartmentalization; that is, they do not have a differentiated nucleus (the DNA is not separated from the cytoplasmic material by a membrane) or organelles (e.g., mitochondria). Further, the two prokaryotic organisms chosen differ in the structure of their cell wall and represent the two most common wall types. *E. coli* is Gram-negative and its cell wall differs from that of *B. megaterium*, which is Gram-positive (for a general description of the characteristics, see Brock and Madigan, 1991). *S. cerevisiae, S. pombe, Y. lipolytica*, and the spores of *P. blakesleeanus* are typical eukaryotic cells that, however, show several differences with respect to one another. Their characteristics under the scanning electron microscope are shown in Figs. 4-7a, 4-7b and 4-7c. *S. cerevisiae* and *Y. lipolytica* differ from *S. pombe* in their mode of division. The former two microorganisms divide by budding (Domínguez et al., 1982; Rodríguez and Domínguez, 1984), whereas *S. pombe* divides by transverse fission (Fig. 4-7d, Mateos and Domínguez, 1991). These yeasts also differ in the composition of their cell-wall polymers, (namely β-glucan, mannan, and chitin in *S. cerevisiae* (Cabib et al., 1982) and *Y. lipolytica* (Vega and Domínguez, 1986) and α-glucan, β-glucan, and galactomannan in *S. pombe* (Mateos and Domínguez, 1991)). Also, the number of chromosomes is different (16 in *S. cerevisiae*; 5 or 6 in *Y. lipolytica*, and 3 in *S. pombe*). Protoplasts are cells lacking a cell wall, the rigid structure that protects them from osmotic changes in the culture medium (Fig. 4-7e). They must therefore be kep in isotonic media in which they adopt a spherical shape. The plasma membrane and the intracellular structures, however, remain intact. Finally, we have analyzed human lymphocytes. these show a similar kind of behavior (unpublished work).

In our opinion, from the dielectrophoretic behavior of this range of microorganisms and cells, two main conclusions can be drawn. The first is that the dielectrophoretic behavior of bacteria, spores, yeasts, and human lymphocytes of different sizes (between 0.1 and 15 μm in diameter), geometric shape, cell composition, mode of cell division, and evolutionary group (prokaryotic and eukaryotic) is similar in nonuniform AC fields in the 0.1–5 MHz frequency range. Our data suggest that to align cells with a view to perform electrofusions, this frequency range must be used. This notion is corroborated by data obtained when the proportionality between the fourth power of the radius and dielectrophoretic yield is compared for a value of 1 MHz (maximum of the frequency spectrum). The results in Table 4-1 show that for spherical and ellipsoid cells (with volumes between 10 and 250 μm^3) the results obtained are similar, although S. *pombe* is an exception because it has a yield higher than that corresponding to its equivalent volume, probably because of its cylindrical shape.

A particularly interesting result is that whereas the frequency spectrum remains constant, the dielectrophoretic yield depends on the metabolic status of the cells. This can be seen in Fig. 4-10. The absolute value of the dielectrophoretic yield tends toward a constant when the cultures reach the stationary phase or when, in the case of S. *cerevisiae*, the organism is grown in a nonfermentable carbon source. This phenomenon is probably related to the energetic state of the cells, reflected in the metabolic

TABLE 4-1. Relationship between the Dielectrophoretic Yield and the Fourth Power of the Radius in Yeast Cells and Spores[a]

Yeast	Y_R^*	a^4	Y_R^*/Y_{RSc}^*	a^4/a_{Sc}^4	δ (%)
Cells					
Saccharomyces cerevisiae	1.90	21.4	1.00	1.00	
Yarrowia lipolytica	2.90	36.0	1.53	1.68	11
Schizosaccharomyces pombe	10.60	61.5	5.57	2.87	
Protoplasts					
Saccharomyces cerevisiae	1.30	17.7	0.68	0.83	22
Spores					
Phycomyces blakesleeanus	14.50	168.0	7.63	7.84	5
Verticillium chlamydosporium	1.32	13.6	0.68	0.63	8
Penicillium expansum	0.57	5.8	0.30	0.27	9

[a] The radius (a) of the cells or spores was determined from the cell volumes calculated with a Coulter counter. To render the measurements comparable, normalized yield (Y_R^*) was calculated by dividing the real dielectrophoretic yield (Y_R) by voltage (V) and cell spore concentration (C): $Y^* = Y_R/VC$. Y_R^*/Y_{RSc}^* and a^4/a_{Sc}^4 represent the dielectrophoretic yield and the fourth power of the radii compared with the values of S. *cerevisiae*. δ represents the deviation between the quotients of the normalized dielectrophoretic yield and those of the fourth power of the radii. All data represent means of four determinations that differed from one another by no more than 10%. Each experiment was carried out a minimum of three times. *Source*: J. Bacteriol., **162**: 790–793, by permission of the American Society for Microbiology.

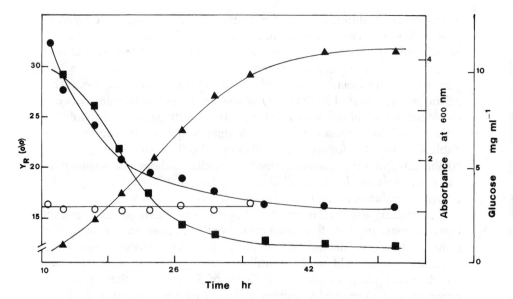

Figure 4-10. Dependence of dielectrophoretic yield (Y_R) on growth and carbon sources of a culture of *Saccharomyces cerevisiae* in minimal medium at 24°C. Cell concentration, 1.9×10^7 cells/ml. Frequency 1 MHz. Electrical conductivity 2×10^{-3} Sm^{-1}. (■) glucose concentration in the medium; (▲) absorbance at 600 nm; (●) yield on glucose (1 and 4%), with or without skaking; (○) yield on ethanol (2%) or acetate (4%), with or without sacking. *Source: J. Bacteriol*, 1985, **162**: 790–793, by permission of the American Society for Microbiology.

activities of the plasma membrane. This should be taken into account in the design of all experiments aimed at obtaining absolute values.

ASSOCIATED PHENOMENA: STIRRING AND PARTICLE REPULSION

When the intensity of the electrical field applied to the electrodes to induce dielectrophoresis is higher than 70000 V/m, turbulent movements and stirring may occur in the suspension medium; that is, the particles in suspension move in directions that do not coincide with the lines of the field. These phenomena are due to thermal convection. Pohl (1978) has shown that increases of a few degrees on the surface of the electrode are sufficient to create convective currents that interfere with dielectrophoresis. Stirring generally occurs in the pin–pin electrode configuration and when the suspension has a high electrical conductivity. Indeed, it may be so strong that it will drag off the dielectrophoretic deposit adhering to the electrodes. Figure 4-11a shows this phenomenon occurring in aqueous

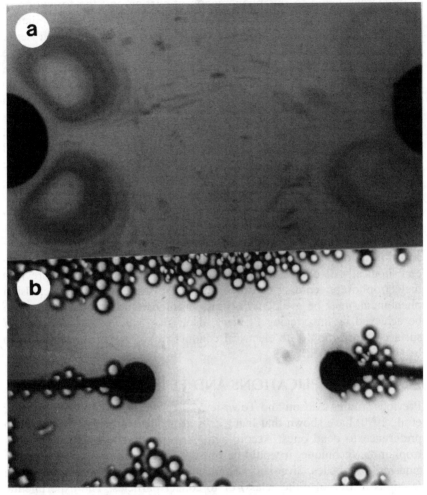

Figure 4-11. Photomicrographs of a) stirring of living *Saccharomyces cerevisiae* cells at 80 V rms and b) negative dielectrophoresis of Sephadex G-50.

suspensions of *S. cerevisiae*. The spinning of the vortices formed around the two electrodes goes in opposite directions to maintain angular momentum. Logically, this stirring phenomenon should be avoided in electrofusion experiments because it prevents stable contact between cells, such that the conductivity of the medium should be controlled carefully. The effect of negative dielectrophoresis (Fig. 4-11b) and aspects of the dielectrophoresis deposit after disconnecting the electrical field (Fig. 4-4c) should also be taken into account.

Intimately linked to this phenomenon of stirring and the formation of vortices is that of the effects of repulsion. The particles that are dielectrophoretically directed toward the metal electrodes change their sense sharply and are repelled by them. Also, the pearl chains adhering to the electrodes may suddenly fragment, and some elements of the chains may, in turn, be repelled from them. This phenomenon is also closely related to the electrical conductivity of the medium. There may also be phenomena of mutual dielectrophoresis among pairs of cells, and, if the dielectrophoretic force is negative, it makes the particles repel each other. Similarly, this repulsion may be due to small localized jets of liquid streaming from the electrodes (Pohl, 1961). The metal surface of the electrodes, in fact, shows microscopic irregularities that may harbor contaminants despite the careful cleaning to which the electrodes are subject. The presence of such contaminants may alter certain electrical properties of the electrodes. This leads to the formation of tiny jets of charge that drag the cells off the electrode. There may also be problems of surface polarization of the electrodes, which also lead to these repulsion phenomena. Logically, if one is attempting to achieve electrofusion by previous dielectrophoretic contact between two cells, these electrical repulsion phenomena must be avoided. For this, an improvement in the system can be achieved by introducing Pt black electrodes that prevent problems of surface polarization. The electrical conductivity of the medium must also be strictly controlled.

PRACTICAL APPLICATIONS AND PERSPECTIVES

Previous studies (Mason and Townsley, 1971; Iglesias et al., 1989; Asencor et al., 1990) have shown that living cells are collected on the electrodes in preference to dead ones. Accordingly, it seems that by using the dielectrophoretic technique, it would be possible to separate different microorganisms, organelles, live and dead cells (or at least to enrich certain fractions in one of either class) or to collect particles. The experimental designs tested until now, however, have not afforded results that can be applied at an industrial scale. Probably, the relationship between the distance at which the electrodes should be placed and the voltage necessary to achieve the adherence of the pearl chains is the limiting aspect of the process.

Dielectrophoretic processes seem to be more promising for measuring parameters of biological interest, such as the membrane capacitance of single cells and the electrical permittivity of the plasma membrane. A further interest of dielectrophoresis is when it is combined with electrorotation (Arnold et al., 1985) and feedback-controlled levitation (Kaler and Jones, 1990). These techniques allow us to analyze not only the behavior of cell suspensions but also that of single cells. Despite this, many experiments have been performed with very different types of cells or protoplasts (e.g., yeasts, plant protoplasts, mammalian cells), in which the metabolic state of the organisms and their origin has not been stated explicitly. Thus, an exact definition or the organism used (e.g., the *Neurospora crassa* cell wallless slime mutant; Fikus et al., 1987) and of the metabolic conditions, active growth, or resting state are essential for obtaining reproducible results.

Dielectrophoresis can also be of use for the study of electroporation. Using dielectrophoresis, it is possible to orient asymmetric cells, thereafter applying the electrical pulse. The applications of electroporation have recently been overviewed by Tsong, 1991. (See Chapters 8, 9, 10 and 11). It is probably electrofusion, however, where the use of dielectrophoresis has the most interesting possibilities from the point of view of biotechnology. Dielectrophoresis can induce cellular contact among cells of the same, or different, types by several mechanisms that have been reviewed by Teissié et al. (1989). Electrofusion can be achieved by applying a dielectrophoretic pulse to concentrated cell suspensions (Fig. 4-12) or by alignment by dielectrophoresis of the pairs of cells to be fused, as has been described by

Figure 4-12. Scanning electron photomicrographs of electrofusionants obtained by mixing intact *Saccharomyces cerevisiae* VY 1160 and VY 2 strains. Electrofusion was carried out in an Electro Cellfusion CFA 500 apparatus (Krüss, GmbH). Pulse voltage, 200 V; number of pulses, 3; pulse duration 10 μs; interval between pulses, 0.5 s. The characteristics of the *S. cerevisiae* fragile mutants have been described by Philipova and Venkov (1990).

Zimmermann, 1982. (See Chapters 1, 5, 6 and 7). With both systems, using suitable methods of selection, it will be possible to obtain fusionants of the correct characteristics and in the necessary amounts for industrial application.

SUMMARY

Dielectrophoresis enables detailed studies to be undertaken of the surface charges and dielectric properties of cells and particles. The technique has application to cell separation, cell electrofusion and in analysis of the orientation of both spherical and non-spherical cells. Microorganisms, especially yeasts with their reproducible growth characteristics, have been employed as biological material in many dielectrophoretic investigations.

ACKNOWLEDGMENTS

We thank N. Skinner for revising the English version of this manuscript. This work was partially supported by grants from the Junta de Castilla y León (Project No SA47/93) and DGICYT (BIO92-0304).

References

Arnold, W. M., Wendt, B., Zimmermann, U., and Korenstein, R. (1985). Rotation of a single swollen thylakoid vesicle in a rotating electric field. Electrical properties of the photosynthetic membrane and their modification by ionophores, lipophilic ions and pH. Biochim. Biophys. Acta **813**: 117–131.

Arnold, W. M., Zimmermann, U., Heiden, W., and Ahlers, J. (1988b). The influence of tetraphenylborates (hydrophobic anions) on yeast cell electrorotation. Biochim. Biophys. Acta **942**: 96–106.

Arnold, W. M., Zimmermann, U., Pauli, W., Benzing, M., Niehrs, C., and Ahlers, J. (1988a). The comparative influence of substituted phenols (especially chlorophenols) on yeast cells assayed by electrorotation and other methods. Biochim. Biophys. Acta **942**: 83–95.

Asencor, F. J., Colom, I., Santamaría, C., Domínguez, A., and Iglesias, F. J. (1990). Comparison between the behaviour of yeast and bacteria in a non-uniform alternating electric field. Bioelectrochem. Bioenerg. **24**: 203–214.

Asencor, F. J., Santamaría, C., Iglesias, F. J., and Domínguez, A. (1993). Dielectric energy of orientation in dead and living cells of Schizosaccharomyces pombe. Fitting of experimental results to a theoretical model. Biophys. J. **64**: 1626–1631.

Benguigui, L., and Lin, I. J. (1982). More about the dielectrophoretic force. J. Appl. Phys. **53**: 1141–1143.

Brock, T. D., and Madigan, M. T. (1991). Biology of Microorganisms. Prentice-Hall Inc., New Jersey, 6th ed.

Cabib, E., Roberts, R., and Bowers, B. (1982). Synthesis of the yeast cell wall and its regulation. Annu. Rev. Biochem. **51**: 763–793.

Crane, J. S., and Pohl, H. A. (1972). Theoretical models of cellular dielectrophoresis. J. Theor. Biol. **37**: 15–41.

Cross, F., Hartwell, L. H., Jackson, C., and Konopka, J. B. (1988). Conjugation in *Saccharomyces cerevisiae*. Annu. Rev. Cell. Biol. **4**: 429–457.

Dimitrov, D. S., Tsoneva, I., Stoicheva, N., and Zhelev, D. V. (1984). An assay for dielectrophoresis: Applications to electromagnetically induced membrane adhesion and fusion. J. Biol. Phys. **12**: 26–30.

Dimitrov, D. S., and Zhelev, D. V. (1987). Dielectrophoresis of individual cells experimental methods and results. Bioelectrochem. Bioenerg. **17**: 549–557.

Domínguez, A., Elorza, M. V., Villanueva, J. R., and Sentandreu, R. (1980). Regulation of chitin synthase activity in *Saccharomyces cerevisiae*: Effect of the inhibition of cell division and of synthesis of RNA and protein. Curr. Microbiol. **3**: 263–266.

Domínguez, A., Varona, R. M., Villanueva, J. R., and Sentandreu, R. (1982). Mutants of *Saccharomyces cerevisiae* cell division cycle defective in cytokinesis: Biosyntesis of the cell wall and morfology. Antonie van Leeuwenhoek J. Microbiol. Serol. **48**: 145–157.

Fielding, G. H., Thompson, J. K., Bogardus, H. F., and Clark, R. C. (1975). Dielectrophoretic filtration of solid and liquid aerosol particulates, 68th Annual Meeting Air Pollution Control Association, Boston.

Fikus, M., Marszalek, P., Rozycki, S., and Zielinsky, J. J. (1987). Dielectrophoresis and electrofusion of *Neurospora crassa* slime. Stud. Biophys. **119**: 73–76.

Förster, E., and Emeis, C. C. (1985). Quantitative studies on the viability of yeast protoplasts following dielectrophoresis. FEMS Microbiol. Lett. **26**: 65–69.

Friend, A. W., Finch, E. D., and Schwan, H. P. (1975). Low frequency electric field induced changes in the shape and motility of Amoebas. Science **187**: 357–359.

Fuhr, G., Glaser, R., and Hagedorn, R. (1986). Rotation of dielectrics in a rotating electric high-frequency field. Biophys. J. **49**: 395–402.

Gimsa, J., Marszalek, P., Loewe, U., and Tsong, T. Y. (1991). Dielectrophoresis and electrofusion of *Neurospora* slime and murine myeloma cells. Biophys. J. **60**: 749–760.

Glaser, R., Fuhr, G., and Gimsa, J. (1983). Rotation of erytrocytes, plant cells, and protoplasts in an outside rotating electric field. Stud. Biophys. **96**: 11–20.

Griffin, J. L., and Stowell, R. E. (1966). Orientation of *Euglena* by radio frequency fields. Exp. Cell Res. **44**: 684–688.

Hatakeyama, T., Taniji, A., and Yagi, H. (1987). Orientation and pearl-chain formation of *Paramecia* induced by A.C. electric field. Jpn. J. Appl. Phys. **26**: 1916–1920.

Hatakeyama, T., and Yagi, H. (1990). A new rotation phenomena of cells induced by homogeneous electric field. Jpn. J. Appl. Phys. **29**: 988–992.

Holzapfel, C., Vienken, J., and Zimmermann, U. (1982). Rotation of cells in an alternating electric field: Theory an experimental proof. J. Membrane Biol. **67**: 13–26.

Iglesias, F. J., López, M. C., Santamaría, C., and Domínguez, A. (1984). Dielectrophoretic properties of yeast cells dividing by budding and by transversal fission. Biochim. Biophys. Acta **804**: 221–229.

Iglesias, F. J., López, M. C., Santamaría, C., and Domínguez, A. (1985). Orientation of *Schizosaccharomyces pombe* nonliving cells under alternating uniform and nonuniform electric fields. Biophys. J. **48**: 721–726.

Iglesias, F. J., Santamaría, C., López, M. C., and Domínguez, A. (1989). Dielectrophoresis. Behavior of microorganisms and effect of electric fields on orientation phenomena. Pages 37–57 in Electroporation and Electrofusion in Cell Biology. Neuman, E., Sowers, A. E., and Jordan, C., eds. Plenum Publishing, New York.

Jackson, C. L., and Hartwell, L. H. (1990a). Courtship in Saccharomyces cerevisiae: An early cell–cell interaction during mating. Mol. Cell. Biol. 5: 2202–2213.

Jackson, C. L., and Hartwell, L. H. (1990b). Courtship in Saccharomyces cerevisiae: Both cell types choose mating partners by responding to the strongest pheromone signal. Cell 63: 1039–1051.

Jackson, C. L., Konopka, J. B., and Hartwell, L. H. (1991). S. Cerevisiae α pheromone receptors activate a novel signal transduction pathway formating partner discrimination. Cell 67: 389–402.

Jones, T. B., and Kallio, G. (1979). Dielectrophoretic levitation of spheres and shells. J. Electrostat. 6: 207–224.

Kaler, K. V., and Jones T. B. (1990). Dielectrophoretic spectra of single cells determined by feedback-controlled levitation. Biophys. J. 57: 173–182.

López, M. C., Iglesias, F. J., Santamaría, C., and Domínguez, A. (1984). Dielectrophoretic behaviour of Saccharomyces cerevisiae: Effect of cations and detergents. FEMS Microbiol. Lett. 24: 149–152.

López, M. C., Iglesias, F. J., Santamaría, C., and Domínguez, A. (1985). Dielectrophoretic behavior of yeast cells: Effect of growth sources and cell wall and a comparison with fungal spores. J. Bacteriol. 162: 790–793.

Marszalek, P., Zielinsky, J. J., Fikus, M., and Tsong, T. Y. (1991). Determination of electric parameters of cell membranes by a dielectrophoretic methods. Biophys. J. 59: 982–987.

Mason, B. D., and Townsley, D. M. (1971). Dielectrophoretic separation of living cells. Can. J. Microbiol. 17: 879–888.

Mateos, P., and Domínguez, A. (1991). Ultrastructure and cell wall composition in cell division cycle mutants of Schizosaccharomyces pombe deficient in septum formation. Antonie van Leeuwenhoek J. Microbiol. Serol. 59: 155–165.

Mischel, M., and Lamprecht, I. (1983). Rotation of cells in nonuniform alternating field. J. Biol. Phys. 11: 43–44.

Mishima, K., and Morimoto, T. (1989). Electric field-induced orientation of myelin figures of phosphatidylcholine. Biochim. Biophys. Acta 985: 351–534.

Philipova, D. H., and Venkov, P. V. (1990). Cell fusion of Saccharomyces cerevisiae fragile mutants. Yeast 6: 205–212.

Pohl, H. A. (1951). The motion and precipitation of suspensoids in divergent electric fields. J. Appl. Phys. 22: 869–871.

Pohl, H. A. (1958). Some effects of nonuniform fields on dielectrics. J. Appl. Phys. 29: 1182–1189.

Pohl, H. A. (1961). Formation of liquid jets in non-uniform electric fields. J. Appl. Phys. 32: 1784–1785.

Pohl, H. A. (1978). Dielectrophoresis. Cambridge University Press, London.

Pohl, H. A., and Crane, J. S. (1971). Dielectrophoresis of cells. Biophys. J. 11: 711–727.

Pollock, J. K., Pohl, D. G., and Pohl, H. A. (1987). Cellular spin resonance inversion by ion-induced dipoles. Int. J. Quantum Chem. 32: 19–29.

Rodríguez, C., and Domínguez, A. (1984). The growth characteristics of *Saccharomycopsis lipolytica*: Morphology and induction of mycelium formation. Can. J. Microbiol. **30**: 605–612.

Saito, M., Schwan, H. P., and Schwarz, G. (1966). Response of nonspherical biological particles to alternating electric fields. Biophys. J. **6**: 313–327.

Santamaría, C., Iglesias, F. J., and Domínguez, A. (1985). Dielectrophoretic deposition in suspensions of macromolecules: Polyvinylchloride and Sephadex G-50. J. Colloid Interface Sci. **103**: 508–515.

Schwarz, G., Saito, M., and Schwan, H. P., (1965). On the orientation of nonspherical particles in an alternating electric field. J. Chem. Phys. **43**: 3562–3569.

Sowers, A. E. (1989). The mechanism of electroporation and electrofusion in erythrocyte membranes. Pages 229–256. In Electroporation and Electrofusion in Cell Biology. Neuman, E., Sowers, A. E., and Jordan, C., eds. Plenum Publishing, New York.

Teissié, J., Rols, M. P., and Blangero, C. (1989). Electrofusion of mammalian cells and giant unilamellar vesicles. Pages 203–214. In Electroporation and Electrofusion in Cell Biology. Neuman, E., Sowers, A. E., and Jordan, C., eds. Plenum Publishing, New York.

Teixeira-Pinto, A. A., Nejelski, L. L. Jr., Cutter, J. L., and Heller, J. H. (1960). The behavior of unicellular organisms in an electromagnetic field. Exp. Cell Res. **20**: 548–564.

Ting, I. P., Jolley, K., Beasley, C. A., and Pohl, H. A. (1971). Dielectrophoresis of chloroplasts. Biochim. Biophys. Acta **234**: 324–329.

Tsong, T. Y. (1991). Electroporation of cell membranes. Biophys. J. **60**: 297–306.

Vega, R., and Domínguez, A. (1986). Cell wall composition of the yeast and mycelial forms of *Yarrowia lipolytica*. Arch. Microbiol. **144**: 124–130.

Verschure, R. H., and Ijlst, L. (1966). Apparatus for continuous dielectric-medium separation of mineral grains. Nature **211**: 619–620.

Vienken, J., Zimmermann, U., Alonso, A., and Chapman, D. (1984). Orientation of sickle red blood cells in an alternating electric field. Naturwissenschaften **71**: 158–159.

Vondrejs, V., Pavlícek, I., Kothera, M., and Palková, Z. (1990). Electrofusion of oriented *Schizosaccharomyces pombe* cells through apical protoplast-protuberances. Biochem. Biophys. Res. Commun. **166**: 113–118.

Zimmermann, U. (1982). Electric field-mediated fusion and related electrical phenomena. Biochim. Biophys. Acta **694**: 227–277.

Zimmeramnn, U., and Arnold, W. M. (1983). The interpretation and use of the rotation of biological cells. Pages 211–221. In Coherent Excitations in Biological Systems. Fröhlich, H., and Kremer, F., eds. Springer-Verlag, Berlin.

Zimmermann, U., Vienken, J., and Pilwat, G. (1981). Rotation of cells in an alternating electric field: The occurrence of a resonance frequency. Z. Naturforsch. **36c**: 173–177.

CHAPTER 5

Electrofusion of Animal Cells

J. A. Lucy

ABSTRACT

This chapter is concerned with the techniques and applications of the electrofusion of animal cells. It draws attention to some of the more important practical factors that need to be considered for success in the technique of electrofusion. Particular attention is given to the various ways of bringing cells in sufficiently close contact for fusion to occur when electrical pulses are applied, since it is this aspect of electrofusion that can most readily be adapted to suit the specific requirements of individual experimentalists. In the second part of the chapter, the success of the method by comparison with the use of poly(ethylene glycol) for the preparation of hybridomas is considered, and recent examples of the practical application of electrofusion in other fields are given.

INTRODUCTION

The electrofusion of animal cells is important for two quite different reasons. First, the technique is a valuable laboratory tool that can be usefully exploited particularly, for example, in the preparation of monoclonal antibodies. Second, work on the molecular mechanisms involved may help us understand the innumerable fusion reactions of biomembranes that occur in animal cell biology at the cellular level, for example, in fertilization and in the fusion of myoblasts, and also at the subcellular level, for example, in exocytosis, and the entry of HIV into lymphocytes. It is not possible, or desirable, however, to treat the techniques of electrofusion separately from its molecular mechanisms because, if electrofusion is to be used to its maximum potential as a laboratory tool, the underlying mechanisms need to be understood. Only then will it be possible to design fully effective fusion protocols for specific experimental purposes.

This chapter is, nevertheless, more concerned with the techniques and applications of the electrofusion of animal cells rather than with molecular mechanisms. Interest in this field is reflected by the publication of a laboratory manual by Borrebaeck and Hagen (1989), which contains descriptions of procedures and equipment used for the production of monoclonal antibodies, and a book by Neumann et al., (1989) in which many aspects of the electrofusion of animal cells, including practical applications, are reviewed.

TECHNIQUES OF ANIMAL CELL ELECTROFUSION

Cell – Cell Contact

Before cells can be fused, they must be brought into close proximity. The repulsive forces between the approaching membranes therefore need to be overcome. For membranes that are 2–3 nm apart, the repulsive forces between phospholipid bilayers are essentially electrostatic. At closer distances, hydration repulsion (arising from solvation of the polar head groups of the phospholipids) dominates the interactions of two approaching membranes (Parsegian et al., 1984; Coorssen and Rand, 1988). High concentrations (30–60% w/v) of poly(ethylene glycol) (PEG), which remove at least 90% of the water from phospholipid bilayers, enable membranes to come extremely close together and ultimately to fuse. Consequently, PEG has been widely used for cell fusion and for the preparation of hybridomas and monoclonal antibodies.

Several different approaches have been adopted to bring animal cells into contact before electrofusion. In most of the early studies, this was done by dielectrophoresis (Arnold and Zimmermann, 1984). Subsequently, in experiments on erythrocyte ghosts, fusion occurred when pulses were applied to cells that were in random positions and that had been brought together after exposure to the breakdown pulses (Sowers, 1989). Teissié and Rols (1986) also showed that electrofusion can be achieved with Chinese hamster ovary cells by centrifuging the cells several minutes after their exposure to electrical field pulses, then incubating the pellet [see also Teissié et al. (1989)]. Apart from the practical issues involved, the fact that cells can be successfully fused when they are brought into contact both before and after electrical breakdown clearly has an important bearing on the molecular mechanisms involved in the fusion process.

Dielectrophoresis

In the presence of an alternating electrical field, dipoles are induced on cells that result in the cells moving in the direction of the greatest field strength when the field is nonuniform. As a consequence, cells that are suspended in a fusion chamber containing two electrodes migrate toward the electrodes. This effect is termed dielectrophoresis (Pohl, 1978) and results in the movement of uncharged particles, whereas electrophoresis applies to charged particles. In dielectrophoresis, many of the migrating cells become attached to one another in so-called "pearl chains." Factors that govern the formation of such pearl chains have been reviewed by Arnold and Zimmermann (1984). Different investigators have used differing AC fields for dielectrophoresis, in part depending on the cell type in question. A suitable voltage for erythrocytes, for example, is 0.4 kV cm^{-1} at a frequency of about 1 MHz. Provided that the applied voltage is kept below about 0l.7 kV cm^{-1}, dielectrophoresis is a fully reversible phenomenon and the pearl chains of cells disperse into random cell suspensions immediately after the AC field is removed. Cell fusion will, however, occur if the AC field exceeds 0.7 kV cm^{-1}, presumably as a consequence of the excessive dielectrophoretic force (Stenger and Hui, 1986).

To minimize the heat generated during dielectrophoresis, cells are usually suspended in essentially nonionic media such as solutions of mannitol, sucrose, or erythritol. Heat generation needs to be avoided as it may damage the cells and prevent fusion from occurring, either by disrupting cell-to-cell contacts in the pearl chains before electrical breakdown or by facilitating resealing of membranes after electrical breakdown (Deuticke and Schwister, 1989). Dielectrophoresis of erythrocyte ghosts may, however, be carried out in weak salt solutions, for example, 60-mM phosphate buffer (Sowers, 1986).

Avidin – Biotin-Mediated Cell Contact

The well-characterized tight binding between avidin and biotin can be used to achieve specific, preselected cell-to-cell contact before the electrofusion of animal cells (Lo et al., 1984). In this procedure, myeloma cells are coated covalently with biotin (or avidin), and the antigen of interest is conjugated covalently to avidin (or biotin). The antigen–avidin conjugate is then used to bring together immunized spleen cells and biotinylated myeloma cells. Application of electrical breakdown pulses to fusion partners that are brought together in this way is reported to produce relatively few hybrid colonies, a very high proportion of which secrete high-affinity antibodies directed specifically against the antigen used. The procedure, therefore, can have an advantage over other, relatively nonspecific, ways of inducing cell fusion, since there is no need to screen very large numbers of the hybridomas that may not be of immediate interest. Some of the most useful monoclonal antibodies have, however, been obtained during such screening processes. Laboratory details of the procedure are provided by Conrad and Lo (1989), but the avidin–biotin method has been used with only a limited number of antigens. Furthermore, as pointed out by Karsten et al. (1988), although the method is elegant, it is neither easy to perform nor applicable to fusions involving cellular antigens.

A simplified method of hybridoma production by avidin-mediated electrofusion, which obviates the requirement for covalent modification of avidin, has been reported by Wojchowski and Sytkowski (1986).

Cell – Cell Contact Induced by Ultrasonic Standing Waves

Cells that are subjected to an ultrasonic standing wave field move to preferred areas in the sound field that are separated by distances of half an acoustic wavelength. Radiation forces also give rise to an intercellular attraction that are at a maximum at these locations (Coakley et al., 1989). The banding of human erythrocytes due to ultrasonic forces is shown in Figure 5-1, in which the cells are concentrated in planes that are at right angles to the direction of sound propagation and are separated from one another by half a wavelength. It is not known how closely individual cells approach each other in such bands. Fusion, however, occurs when aggregated cells in a megahertz frequency standing wave are exposed to a DC electrical breakdown pulse. This phenomenon has been termed electroacoustic fusion, and it has been applied to human erythrocytes and to myeloma cells (Vienken et al., 1985).

Bardsley et al. (1989) have shown that electroacoustic cell fusion can be used with electrically conducting solutions; that is, in physiological media, and that fusion chambers can be used that contain milliliter volumes of cell suspension. These workers have also determined the conditions for the preparation of murine hybridomas by electroacoustic cell fusion, explored the viability of the fusion products, and compared the technique with fusion involving dielectrophoresis (Bardsley et al., 1990). Under optimum conditions, the yield of hybridomas formed by the electroacoustic of cells suspended in mannitol solutions was at least as good as that obtained when the cells were brought in contact by dielectrophoresis. The electro-

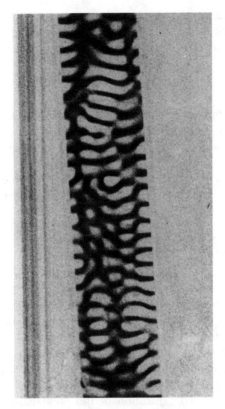

Figure 5-1. The aggregation of human erythrocytes into bands, half a wavelength apart and at right angles to the direction of sound propagation, induced by exposure of the cells to an ultrasonic standing wave. Reproduced with permission from Vienken et al., (1985).

acoustic fusion yield was also comparable to conventional yields when the cells were exposed to a pulse in higher ionic strength media, which give rise to heating effects in dielectrophoresis.

ELECTRICAL BREAKDOWN AND CELL FUSION

Electrical Parameters

The exposure of animal cells to between one and three very short electrical pulses (each of about 10–20 μsec duration at 1 sec intervals), at a voltage of some 2–5 kV cm^{-1}, results in a temporary breakdown of the plasma membrane, which subsequently regains its high electrical resistance and impermeability. For spherical cells, the induced membrane voltage (V_m) at any point on the cell surface is given approximately by the equation

$$V_m = 1.5 \cdot E \cdot \alpha \cdot \cos \Theta$$

where E is the applied field strength, α is the cell radius in centimeters, and Θ is the angle between the radius from the point on the cell surface and the direction of the field (Arnold and Zimmermann, 1984). It is apparent from this equation that the voltage needed to break down the plasma membrane of a cell is reached at lower values of applied field strength than the breakdown voltage of its intracellular organelles, which necessarily have a smaller radius.

Membrane Phospholipids

Cell fusion induced by electrical field pulses may involve changes in the organization of the phospholipids in the plasma membrane of the permeabilized cells. A new anisotropic peak with respect to control cells has been observed on ^{31}P NMR spectroscopic analysis of the phospholipid components of hamster ovary cells after permeabilization by electrical field pulses (Lopez et al., 1988). This peak, which was very different from that seen with hexagonal phase phospholipids, was present only while the cells were permeable; normal anisotropy was recovered after resealing. It was proposed that a reorganization of the polar head group region that leads to a weakening of the hydration layer may account for the observations made (Lopez et al., 1988).

In earlier work with erythrocytes, the almost exclusive localization of phosphatidylethanolamine in the inner leaflet was completely lost in ghosts prepared by colloid-osmotic lysis after electrical breakdown and resealing. Phosphatidylserine was much less affected, but it was suggested that an enhanced mobility of phosphatidylethanolamine may facilitate the merging of two closely apposed bilayers (Dressler et al., 1983). It is relevant that, in later work, the swelling and osmotically induced fusion of human erythrocytes in the presence of Ca^{2+} has been observed to be associated with the exposure of phosphatidylserine in the outer leaflet of the plasma membrane (Baldwin et al., 1990).

Skeletal Reorganization

In the electrofusion of animal cells, the breakdown pulses presumably modify the membrane skeleton in the case of erythrocytes and, possibly, the cytoskeleton with other cell types, in addition to causing changes in the phospholipid bilayer of the plasma membrane. Intermembrane protein exchange was found by Donath and Arndt (1984) to have a surprisingly high diffusion coefficient (3×10^{-9} cm^2/sec) in human erythrocytes subjected to electrofusion, and it was concluded that this was due to a diminished interaction with the plasma membrane. It has also been reported that the incidence of electrofusion of human erythrocytes increases sharply when the proteins of the membrane skeleton are denatured by heating the cells to 50°C (Glaser and Donath, 1987). With Chinese hamster ovary cells, prepulse incubation with colchicine has been observed to increase the fusion yield slightly (Blangero et al., 1989). In addition, microtubules were found to disappear during the first few minutes after the pulse and reform on subsequent incubation. It was suggested that depolymerization of microtubules may be a consequence of the loss of small metabolites consequent on cell permeabilization by the breakdown pulse and that repolymerization of microtubules occurs when the plasma membrane recovers its barrier character.

FACTORS AFFECTING THE YIELD OF FUSED CELLS

Pronase, Ca^{2+}, Mg^{2+}, and pH

In early work, it was discovered that treatment of both murine virus-induced leukemic cells (Friend cells) and sea urchin eggs with the proteolytic enzyme, pronase (or dispase), stabilized the cells against damaging effects of an electrical breakdown pulse (Zimmerman et al., 1981; Pilwat et al., 1981). Cells treated in this way suffered no adverse effects when exposed to field strengths that would normally lead to irreversible changes in the plasma membrane. In these experiments, it was also observed that when the cells were washed several times with pronase-free, nonelectrolyte solution they again became "sensitive" to the field pulses. Pronase was later used in the fusion of murine (Vienken and Zimmermann, 1982) and human (Bischoff et al., 1982) lymphocytes with myeloma cells to yield hybridoma cells.

Subsequent work has indicated that commercial preparations of pronase have two effects on the electrofusion of animal cells. First, it was observed that, when mouse lymphoma cells (L5178Y) were suspended in a nominally Ca^{2+}-free medium, more than 60% of the cells were irreversibly damaged by the field pulses, and less than 10% of the cells were fused. By contrast, relatively high fusion yields (up to 30%) and cell viability (up to 70%) were found in the presence of 30 μM Ca^{2+} (Ohno-Shosaku and Okada, 1984).

Since a solution (1 mg mL^{-1}) of a commercial preparation of pronase was found to contain about 0.1 mM Ca^{2+}, it was therefore suggested that the contaminating Ca^{2+} ions in pronase facilitate electrofusion (Ohno-Shosaku and Okada, 1985). Washing the pronase-treated cells before exposure to the breakdown pulse may well reverse this effect. Millimolar concentrations of Mg^{2+} also stabilize cells against the disruptive effects of electrical field pulses (Ohno-Shosaku and Okada, 1985), and Mg^{2+} ions are included in commercially available media for the fusion of animal cells.

Since heat-inactivated pronase is no longer active and protease inhibitors suppress the effect of the enzyme, it also appears that part of the increase in electrofusion yield found on treating cells with pronase or other proteolytic enzymes is attributable to their proteolytic action on the cell surface (Ohno-Shosaku and Okada, 1984, 1985). This interpretation is supported by the finding that the effects of pronase and of dispase on L5178Y cells are not abolished by removing the enzymes before exposing the cells to the electrical field pulses (Ohno-Shosaku and Okada, 1989).

The effect of pH on the electrofusion of human erythrocytes, aligned by dielectrophoresis, has been investigated by monitoring the movement from labeled to unlabeled cells of the carbocyanine probe, DiI. The optimum pH was around 7.5. Interestingly, the fusion yield decreased by 40% when the pH was changed from 7.5 to 6.0, but there was only a 20% decrease in yield between pH 7.5 and 10.0 (Chang et al., 1989).

Heavy Metals, Monovalent Ions, Ethanol and Macromolecules

Molecules and ions that would normally have relatively little toxic effect on cells because they are excluded by the plasma membrane can cause cell death if they are present when cells are permeabilized during the electrofusion process. Therefore, it is important to ensure that fusion media are free from pH indicators, such as phenol red. In addition, trace concentrations of metal ions should be removed by shaking the solutions with Chelex beads (BioRad, Richmond, CA; Zimmermann et al., 1989). Interestingly, this contrasts with a recently reported finding that the ability of PEG to induce cell fusion appears to depend on the presence of metal ions in commercial preparations of the polymer. PEG that is treated with Chelex beads is unable to induce the fusion of human erythrocytes, and its fusogenic properties can be restored by the addition of trace quantities of bivalent or trivalent metal ions (Ahkong and Lucy, 1990). The ionic composition of the medium has been observed to influence the electrofusion yield with Chinese hamster ovary cells. Thus, increasing the concentration of monovalent cations decreased the fusion yield in the order Li$^+$ < choline < Na$^+$ = K$^+$, without affecting cell viability (Blangero and

Teissié, 1985). Fusion did not occur when the medium contained 50 mM NaCl or KCl. Incubation of these cells with ethanol before field pulsation has also been found to decrease the fusion yield; it was suggested that this was due to a decrease in size of the field-induced pores that results from an increase in membrane fluidity (Orgamide et al., 1985). In a subsequent study, the action of ethanol has been considered in relation to four successive steps in the electropermeabilization of membranes; namely, the induction of temporarily permeable structures, their expansion, stabilization, and resealing. It was concluded that ethanol and lysolecithin (which conversely facilitates electrofusion), affect only the expansion and resealing steps (Rols et al., 1990).

It has been found that the presence in ghosts prepared from rabbit erythrocytes of various amounts of haemoglobin, bovine serum albumin, or dextran at low concentrations result in fusion yields being significantly higher than when the ghosts contained buffer only (Sowers, 1990). The fusion yield was also affected by small changes in the concentration of haemoglobin when it was present outside the ghosts' membranes in the suspension buffer.

Osmolarity

In experiments undertaken in this laboratory on the effects of osmolarity on the efficiency of electrofusion, human erythrocytes were labeled with carboxyfluorescein, and a mixture of labeled and unlabeled cells were then suspended in a hypotonic solution of 150-mM erythritol and 10-mM histidine, containing pronase (1 mg ml^{-1}). On exposure to a field pulse, unlabeled erythrocytes that were adjacent to labeled cells became fluorescent, and pearl chains containing up to 14 fluorescent cells were formed (Ahkong et al., 1986). By contrast, the fluorophore transferred only to single cells that were immediately adjacent to the fluorescent cells when the erythrocytes were suspended in 200-mM erythritol, and few fused cells were then formed. In 400-mM erythritol, the instantaneous transfer of fluorescence was negligible and fused cells were not formed. In subsequent work, NS1 mouse myeloma cells were found to behave in a similar manner, and it was suggested that hypotonic media may thus be of value in the fusion of myeloma cells with lymphocytes for the preparation of monoclonal antibodies (Brown et al., 1986).

In their studies on the electroacoustic fusion of human erythrocytes, Bardsley et al., (1989) have since found that the fusion yield in 170 mOsm solutions is much higher than the yield from cells in 272 mOsm solutions. With cells aligned by dielectrophoresis, Schmitt and Zimmermann (1989) have also observed that the electrofusion of mammalian cells in strongly hypoosmolar media containing sorbitol, small amounts of divalent cations and albumin result in high yields of hybridoma cells. Optimum clone numbers of hybrid cells were obtained for the fusion of osmotically stable subclones of murine myeloma cells with lymphocytes (stimulated with DNP-haemocyanin) provided that the osmolarity of the fusion medium was as low as 75 mOsm. The efficacy of the hypoosmolar electrofusion of these cells allowed the use of very few cells (about 10^5 lymphocytes per fusion chamber). Details of the application of hypoosmolar fusion media are described by Zimmermann et al. (1989).

Pulsed Radio-Frequency Electric Fields

As emphasized earlier, it is difficult to fuse cells of different sizes because large cells are more sensitive to electrical breakdown than small cells. A possible solution to this problem may lie in the development of cell fusion by pulsed radio-frequency electrical fields, which permits a reversible breakdown of the plasma membranes of two cells of different size without irreversibly damaging the larger cell (Chang, 1989a, 1989b). In this technique, cells are first aligned in pearl chains by dielectrophoresis, and then exposed to high-power radio-frequency pulses. As with direct current breakdown pulses, neighboring cells begin to fuse immediately after the oscillating pulses are applied. Use of an oscillating field can, however, sometimes give a fusion yield in excess of 80% with human erythrocytes (Chang, 1989b). A further possible advantage of this technique may consequently lie in the possibility of improving the fusion yield of hybridoma cells.

APPLICATIONS

Hybridomas and Monoclonal Antibodies

The preparation of hybridoma cells by electrofusion and the production of monoclonal antibodies have excited much interest, possibly because of the potential commercial value of human monoclonal antibodies in the clinical field. Refer to the review by Perkins et al., (1989) for a general account of the practicalities of the preparation of hybridomas that secrete human monoclonal antibodies (based on the application of electrical field pulses to cells aligned by dielectrophoresis).

PEG has disadvantages for preparing hybridomas, which have been succinctly summarized by Karsten et al. (1988). These investigators have reported a direct comparison of electrofusion and PEG-mediated fusion for the generation of antibody-producing hybridomas, in experiments with several different types of antigens. Compared with PEG, electrofusion resulted in a 3.8–33.0 times higher yield of hybridomas per unit number of spleen cells, which was statistically significant at the 5% level and independent of the type of antigen. Hybridomas from electrofusion initially grew faster and were visible earlier. After cloning, hybridomas arising from the two groups were apparently indistinguishable in every respect; for example, antibody titre, distribution of isotypes, stability of antibody production, transplantability, and cryopreservation.

Ohnishi et al., (1987) have confirmed that the presence of both Ca^{2+} and Mg^{2+} in the fusion medium and pretreatment of the mixed cells with proteases gives an improved hybridoma yield. The presence of the chemical fusogen, glyceryl monooleate (Ahkong et al., 1973), and of calmodulin increased the hybridoma yield about eightfold. Phosphatidylserine also facilitated the yield slightly. Under optimum conditions, hybridoma yields were about 10 times higher than with PEG. Electrofusion has also been found to give a higher fusion efficiency in comparison with PEG-induced fusion by Foung and Perkins (1989) in experiments with B cells activated by Epstein–Barr virus or pokeweed mitogen. They used the technique to produce a panel of monoclonal antibodies to human cytomegalovirus. Katenkamp et al. (1989) have also reported that, by comparison with the use of PEG, electrofusion gives a distinctly higher yield of hybridoma cells under optimum conditions.

As discussed in the section on osmolarity, hypoosmolar solutions have been found to enhance the yield of hybridoma cells (Schmitt and Zimmermann, 1989; Zimmermann et al., 1989). A novel 20-μL fusion chamber has been described by Glaser et al. (1989), which enables cell samples to be submitted to electrofusion every 2 min and allows the fusion process to be monitored with a microscope. They observed that the use of pronase, and the establishment of normal ionic strength a short time after electrical breakdown, were essential to obtain up to 15 hybridoma clones per 100,000 lymphocytes.

Comparative experiments on the production of human cytotoxic T-cell hybridomas by PEG and electrofusion, preceded by dielectrophoresis, have been described by Gravekamp et al. (1987). Cytotoxic human T cells were fused with human T-lymphoma cells and mouse B-myeloma cells. To reduce the cytotoxic activity of the T lymphocytes during the fusion procedure, they were heated to 46°C for 10 min. For one tumor cell line, considerably more hybridomas were obtained by electrofusion than with PEG (with or without heat shock). Only 4, in more than 600 hybridomas tested, however showed transient cytotoxic activity, but in none of them was this function immortalized. It was therefore concluded that other strategies, such as a DNA-mediated transfer of genes that endow human cytotoxic T cells with unlimited cytotoxic proliferative capacity, may be more promising.

Examples of Other Applications

A number of studies on the electrofusion of mammalian gametes has been reported. For example, Kubiak and Tarkowski (1985) showed that the fusion of two-cell mouse embryos having an intact zona pellucida can be induced with electrical pulses and that the fused blastomeres develop into tetraploid blastocysts, but died after implantation. Exposure to an AC field before application of the DC breakdown pulses did not increase the fusion yield. It was concluded that, in cleaving embryos with an intact zona pellucida, the natural contact between sister blastomeres is sufficiently strong for the fusion to be achieved by DC pulses only. Whereas Kubiak and Tarkowski used an electrolyte solution, Winkel and Nuccitelli (1989) found that a nonelectrolyte medium reduced lysis of mouse embryos. They developed a procedure for the formation of octaploid blastomeres from four-cell mouse embryos, which were fused by exposure to a DC field of more than 2000 V/cm for 10 sec. The octaploid embryos divided normally and were observed to cavitate and polarize at the same time as normal embryos, indicating that the mechanisms that underlie cell division, cavitation, and cortical polarization are not affected by changes in cell size or ploidy. Sun and Moor (1989) disaggregated zona-free two-cell mouse embryos into single blastomeres and aligned them by dielectrophoresis before electrofusion. High fusion rates were the product of an inverse relationship between DC field strength and pulse duration, but the initiation of pore formation was insufficient to induce successful fusion unless accompanied by an appropriate post-pulse medium and adequate membrane contact.

With the aim of constructing an animal model for *Neisseria gonorrhoea* (which binds exclusively to receptors in the plasma membranes of human cells), Grasso et al. (1989) and Heller and Grasso (1990) have fused human U937 lymphoma cells to rabbit corneal epithelial tissue both in vitro and in situ. They then showed that the bacteria will attach to the modified corneal interfaces. This indicates that gonococcal membrane receptors from human cells can be functionally incorporated into superficial rabbit corneal epithelia and that it is feasible to use electrofusion techniques to establish novel animal models. Interestingly, these workers also demonstrated that tissue damage due to heat produced by AC electrophoresis could be circumvented by using mechanical pressure to achieve close cell-tissue juxtaposition before electrofusion.

FUTURE PROSPECTS

Future technical advances in the electrofusion of animal cells may arise from attention to the most appropriate way of inducing cell–cell contact in specific experimental situations, particularly as fusion can be achieved relatively simply by using mechanical pressure on adjacent cells before electrical breakdown or by centrifuging cells several minutes after exposure to electrical field pulses (Teissié and Rols, 1986; Teissié et al., 1989). Electroacoustic fusion, although more complex, may also be advantageously exploited in this respect.

Some of the most important future applications of the electrofusion of animal cells may well lie in its ability to induce fusion in situ, such as in the corneal epithelia studied by Grasso et al. (1989). Another field in which advances might be expected from the application of electrofusion is in the preparation of T-cell hybridomas, particularly human T–T cell hybridomas (Fox and Platsoucas, 1990), which has received relatively little attention by comparison with investigations on B-cell hybridomas.

A new development of much promise, which is closely related to the fusion of animal cells, is electroinsertion. This consists of the application of pulsed, microsecond electrical fields to a suspension of cells in the presence of a membrane protein having a membrane-spanning region (Mouneimne et al., 1990). Electroinsertion has been used to insert the full-length recombinant CD4 receptor into human and murine erythrocyte membranes, without damage to the cells but with a substantial fraction of the molecules properly exposed and immunologically active. By contrast, attempts to electroinsert proteins without a membrane-spanning sequence consistently failed.

References

Ahkong, Q. F., Fisher, D., Tampion, W. and Lucy, J. A. (1973). The fusion of erythrocytes by fatty acids, esters, retinol and α-tocopherol. Biochem. J. **136**: 147–155.

Ahkong, Q. F., and Lucy, J. A. (1986). Osmotic forces in artificially induced cell fusion. Biochim. Biophys. Acta **858**: 206–216.

Ahkong, Q. F., and Lucy, J. A. (1990). Chelex-100 treated polyethylene glycol is non fusogenic. Proc. 10th Int. Biophysics Congress. Abstract P3.5.10.

Arnold, W. M., and Zimmermann, U. (1984). Electric field-induced fusion and rotation of cells. Pages 389–454. In Biological Membranes, Vol. 5, Chapman, D. ed., Academic Press, London.

Baldwin, J. M., O'Reilly, R. Whitney, M., and Lucy J. A. (1990). Surface exposure of phosphatidylserine is associated with the swelling and osmotically-induced fusion of human erythrocytes in the presence of Ca^{2+}. Biochim. Biophys. Acta **1028**: 14–20.

Bardsley, D. W., Coakley, W. T., Jones, G., and Liddell, J. E. (1989). Electroacoustic fusion of millilitre volumes of cells in physiological media. J. Biochem. Biophys. Meths. **19**: 339–348.

Bardsley, D. W., Liddell, J. E., Coakley, W. T., and Clarke, D. J. (1990). Electroacoustic production of murine hybridomas. J. Immunol. Methods, **129**: 41–47.

Bischoff, R., Eisert, R. M., Schedel, I., Vienken, J., and Zimmermann, U. (1982). Human hybridoma cells produced by electro-fusion. FEBS Lett. **147**: 64–68.

Blangero, C., and Teissié, J. (1985). Ionic modulation of electrically induced fusion of mammalian cells. J. Membrane Biol. **86**: 247–253.

Blangero, C., Rols, M. P., and Teissié, J. (1989). Cytoskeletal reorganization during electric-field-induced fusion of Chinese hamster ovary cells grown in monolayers., Biochim. Biophys. Acta **981**: 295–302.

Borrebaeck, C. A. K., and Hagen, I. (1989). Electromanipulation in Hybridoma Technology. Stockton Press, New York.

Brown, S. M., Ahkong, Q. F., and Lucy, J. A. (1986). Osmotic pressure and the electrofusion of myeloma cells. Biochim. Soc. Trans. **14**: 1129–1130.

Chang, D. C. (1989a). Cell fusion and cell poration by pulsed radio-frequency electric fields. Pages 215–227. In Electroporation and Electrofusion in Cell Biology. Neumann, E., Sowers, A.E., and Jordan, C.A., eds. Plenum Press, New York.

Chang, D. C. (1989b). Cell poration and cell fusion using an oscillating electric field. Biophys. J. **56**: 641–652.

Chang, D. C., Hunt, J. R., and Gao, P. (1989). Effects of pH on cell fusion induced by electric fields. Cell Biophysics **14**: 231–243.

Coakley, W. T., Bardsley, D. W., and Grundy, M. A. (1989). Cell manipulation in ultrasonic standing wave fields. J. Chem. Tech. Biotechnol. **44**: 43–62.

Conrad, M. K., and Lo, M. M. S. (1989). B-cell hybridoma production by avidin-biotin mediated electrofusion. Pages 89–102. In Electromanipulation in Hybridoma Technology. Borrebaeck, C.A.K., and I. Hagen, I. eds. Stockton Press, New York.

Coorssen, J., and Rand, R. P. (1988). Competitive forces between lipid membranes. Studia Biophysica **127**: 53–60.

Deuticke, B., and Schwister, K. (1989). Leaks induced by electrical breakdown in the erythrocyte membrane. Pages 127–148. In Electroporation and Electrofusion in Cell Biology. Neumann, E., Sowers, A.E., and Jordan, C.A., eds. Plenum Press, New York.

Donath, E., and Arndt, R. (1984). Electric-field-induced fusion of enzyme-treated human red cells: Kinetics of intermembrane protein exchange. Gen. Physiol. Biophys. **3**: 239–249.

Dressler, V., Schwister, K., Haest, C. W. M., and Deuticke, B. (1983). Dielectric breakdown of the erythrocyte membrane enhances transbilayer mobility of phospholipids. Biochim. Biophys. Acta **732**: 304–307.

Foung, S. K. H., and Perkins, S. (1989). Electric field-induced cell fusion and human monoclonal antibodies. J. Immunol. Methods **116**: 117–122.

Fox, F. E., and Platsoucas, C. D. (1990). Human T-T cell hybridomas: Development and applications. Hum. Antibod. Hybridomas **1**: 3–9.

Glaser, R. W., and Donath, E. (1987). Hindrance of red cell electrofusion by the cytoskeleton. Studia Biophysica **121**: 37–43.

Glaser, R. W., Jahn, S., and Grunow, R. (1989). Electrofusion of human B-cell hybridomas. Influence of some physical and chemical factors on hybridoma yield. Studia Biophysica **130**: 201–204.

Grasso, R. J., Heller, R., Cooley, J. C., and Haller, E. M. (1989). Electrofusion of individual animal cells directly to intact corneal tissue. Biochim. Biophys. Acta **980**: 9–14.

Gravekamp, C., Santoli, D., Vreugdenhil, R., Collard, J. G., and Bolhuis, R.L.H. (1987). Efforts to produce human cytotoxic T-cell hybridomas by electrofusion and PEG fusion. Hybridoma **6**: 121–133.

Heller, R., and Grasso, R. J. (1990). Transfer of human membrane surface components by incorporating human cells into intact animal tissue by cell-tissue electrofusion in vivo. Biochim. Biophys. Acta **1024**: 185–188.

Karsten, U., Stolley, P., Walther, I., Papsdorf, G., Weber, S., Conrad, K., Pasternak, L., and Kopp, J. (1988). Direct comparison of electric field-mediated and PEG-mediated cell fusion for the generation of antibody producing hybridomas. Hybridoma **7**: 627–633.

Katenkamp, U., Schumann, I., and Wolf, I. (1989). Contributions to the use of electrofusion for generation of hybridoma cells. Studia Biophysica **130**: 205–210.

Kubiak, J. C., and Tarkowski, A. K. (1985). Electrofusion of mouse blastomeres. Exp. Cell Res. **157**: 561–566.

Lo, M. M. S., Tsong, Y. T., Conrad, M. K., Strittmatter, S. M., Hester, L. H., and Snyder, S. H. (1984). Monoclonal antibody production by receptor-mediated electrically-induced cell fusion. Nature **310**: 792–794.

Lopez, A., Rols, M. P., and Teissié, J. (1988). ^{31}P NMR analysis of membrane phospholipid organization in viable, reversible, electropermeabilised Chinese hamster ovary cells. Biochemistry **27**: 1222–1228.

Mouneimne, Y., Tosi, P-F., Barhoumi, R., and Nicolau, C. (1990). Electroinsertion of full length recombinant CD4 into red blood cell membrane. Biochim. Biophys. Acta **1027**: 53–58.

Neumann, E., Sowers, A. E., and Jordan, C. A. (1989). Electroporation and Electrofusion in Cell Biology. Plenum Press, New York.

Ohnishi, K., Chiba, J., Goto, Y., and Tokunaga, T. (1987). Improvement in the basic technology of electrofusion for generation of antibody-producing hybridomas. J. Immunol. Methods **100**: 181–189.

Ohno-Shosaku, T., and Okada, Y. (1984). Facilitation of electrofusion of mouse lymphoma cells by the proteolytic action of proteases. Biochem. Biophys. Res. Comm. **120**: 138–143.

Ohno-Shosaku, T., and Okada, Y. (1985). Electric pulse-induced fusion of mouse lymphoma cells: Roles of divalent cations and membrane lipid domains. J. Membrane Biol. **85**: 269–280.

Ohno-Shosaku, T., and Okada, Y. (1989). Role of proteases in electrofusion of mammalian cells. Pages 193–202. In Electroporation and Electrofusion in Cell Biology. Neumann, E., Sowers, A.E., and Jordan, C.A., eds. Plenum Press, New York.

Orgamide, G., Blangero, C., and Teissié, J. (1985). Electrofusion of Chinese hamster ovary cells after ethanol incubation. Biochim. Biophys. Acta **820**: 58–62.

Parsegian, V. A., Rand, R. P., and Gingell, D. (1984). Lessons for the study of membrane fusion from membrane interactions in phospholipid systems. Pages 9–27. In Cell Fusion. Evered, D., and Whelan, J., eds. Pitman, London.

Perkins, S., Zimmermann, U., Gessner, P., and Foung, S. K. H. (1989). Formation of hybridomas secreting human monoclonal antibodies with mouse–human fusion partners. Pages 47–70. In Electromanipulation in Hybridoma Technology. Borrebaeck, C.A.K., and Hagen, I., eds. Stockton Press, New York.

Pilwat, G., Richter, H. P., and Zimmermann, U. (1981). Giant culture cells by electric field-induced fusion. FEBS Lett. 133: 169–174.

Pohl, H. A. (1978). Dielectrophoresis. The behaviour of neutral matter in nonuniform electric fields. Cambridge University Press, Cambridge.

Rols, M. P., Dahhou, F., Mishra, K. P., and Teissié, J. (1990). Control of electric field induced cell membrane permeabilization by membrane order. Biochemistry 29: 2960–2966.

Schmitt, J. J., and Zimmermann, U. (1989). Enhanced hybridoma production by electrofusion in strongly hypo-osmolar solutions. Biochim. Biophys. Acta 983: 42–50.

Sowers, A. E. (1986). A long-lived fusogenic state is induced in erythrocyte ghosts by electric pulses. J. Cell Biol. 102: 1358–1362.

Sowers, A. E. (1989). The mechanism of electroporation and electrofusion in erythrocyte membranes. Pages 229–256. In Electroporation and Electrofusion in Cell Biology. Neumann, E., Sowers, A. E., and Jordan, C.A., eds. Plenum Press, New York.

Sowers, A. E. (1990). Low concentrations of macromolecular solutes significantly affect electrofusion yield in erythrocyte ghosts. Biochim. Biophys. Acta 1025: 247–251.

Stenger, D. A., and Hui, S. W. (1986). Kinetics of ultrastructural changes during electrically induced fusion of human erythrocytes. J. Membrane Biol. 93: 43–53.

Sun, F. Z., and Moor, R. M. (1989). Bioelectrochemistry and Bioenergetics 21: 149–160.

Teissié, J., and Rols, M. P. (1986). Fusion of mammalian cells in culture is obtained by creating the contact between cells after their electropermeabilization. Biochem. Biophys. Res. Comm. 140: 258–266.

Teissié, J., Rols, M. P., and Blangero, C. (1989). Electrofusion of mammalian cells and giant unilamellar vesicles. Pages 203–214. In Electroporation and Electrofusion in Cell Biology. Neumann, E., Sowers, A. E., and Jordan, C.A., eds. Plenum Press, New York.

Vienken, J., and Zimmermann, U. (1982). Electric field-induced fusion: Electrohydraulic procedure for production of heterokaryon cells in high yield. FEBS Lett. 137: 11–13.

Vienken, J., Zimmermann, U., Zenner, H. P., Coakley, W. T., and Gould, R. K. (1985). Electro-acoustic fusion of erythrocytes and of myeloma cells. Biochim. Biophys. Acta 820: 259–264.

Winkel, G. K., and Nuccitelli, R. (1989). Octaploid mouse embryos produced by electrofusion polarize and cavitate at the same time as normal embryos. Gamete Res. 23: 93–107.

Wojchowski, D. M., and Sytkowski, A. J. (1986). Hybridoma production by simplified avidin-mediated electrofusion. J. Immunol. Methods 90: 173–177.

Zimmermann, U. (1986). Electrical breakdown, electropermeabilization and electrofusion. Rev. Physiol. Biochem. Pharmacol. **105**: 175–256.

Zimmermann, U., Gessner, P., Wander, M., and Foung, S. K. H. (1989). Electroinjection and electrofusion in hypo-osmolar solution. Pages 1–30. In Electromanipulation in Hybridoma Technology. Borrebaeck, C. A. K., and Hagen, I., eds. Stock Press, New York.

Zimmermann, U., Pilwat, G., and Richter, H-P. (1981). Electric-field-stimulated fusion: Increased field stability of cells induced by pronase. Naturwissenschaften. **68**: S.577.

CHAPTER 6

Cell-Tissue Electrofusion

R. Heller

M. Jaroszeski

ABSTRACT

Cell–tissue electrofusion is an electromechanical process by which individual, separated animal cells can be electrofused directly to histologically intact tissues in vitro and in vivo. The formation of somatic cell hybrids in vivo is accomplished by applying electrical fields that result in the coalescence of juxtaposed plasma membranes of the individual cells and the cells within the tissues. During the past several years, many protocols have been developed that facilitate incorporation of individual cells into intact tissue. The procedures have been used for the interspecies transfer of specific membrane surface components from human cells to rabbit corneal epithelium. It was also determined that the individual cells were completely incorporated into the tissue within 20 min. A spectrofluorometric assay was used to determine the number of cells fused and to examine parameters involved in the fusion process. The results obtained thus far suggest several potential applications of cell–tissue electrofusion, including site-specific delivery systems, improved surgical procedures, and the establishment of unique animal models.

INTRODUCTION

Electrofusion is a process by which membrane fusion can be induced by exposing juxtaposed cell membranes to electrical fields (Bates et al., 1987;

119

Zimmermann, 1982). Senda et al. (1979) were the first to describe the fusion of two cells by electrofusion. Plant protoplasts were pushed together with micropipettes and fused with a single electrical pulse. Subsequently, other investigators reported similar results with animal (Neumann et al., 1980; Richter et al., 1981) and yeast cells (Weber et al., 1981a, 1981b). Electrofusion has many practical applications, such as the formation of hybridomas (Glassy, 1988; Hewish and Werkmeister, 1989; White et al., 1989; Kwekkeboom et al., 1992), the production of monoclonal antibodies (Klock et al., 1992; Foung and Perkins, 1989; Lo et al., 1984; Lo and Tsong, 1989), studying membrane fusion mechanisms (Sowers, 1987, 1989a, 1989b; Abidor and Sowers, 1992), and examining cytosolic events (Prudovsky and Tsong, 1991; Sixou and Teissie, 1992; Ozawa et al., 1985). In addition, electrofusion has proved to be a valuable tool in examining membrane interactions between two cells or within a single cell (Chernomordik et al., 1987; Frederik et al., 1989; Oberleithner, 1991; Sowers, 1990). Cell–tissue electrofusion (CTE) represents an area of electrofusion that can be used to incorporate individual cells into intact tissue (Grasso et al., 1989). Most of these procedures have been worked out over the past several years.

The incorporation of cells into tissue is accomplished using cell–tissue electrofusion technology. CTE uses electromechanical processes by which individual, separated animal cells can be electrofused directly to histologically intact tissues in vitro or in vivo in anesthetized animals (Grasso et al., 1989; Heller and Grasso, 1990). CTE is achieved by applying electrical fields that result in the coalescence of juxtaposed plasma membranes of individual cells and cells within the tissues. Protocols were developed to incorporate individual human cells into intact rabbit corneal epithelium. Human membrane surface components present on individual cells were incorporated and remained functional on the newly formed human–rabbit somatic cell hybrids. This was demonstrated by the ability of the obligate human pathogen *Neisseria gonorrhoeae* to attach to the histologically modified rabbit tissue. Thus, cell–tissue electrofusion can be used to create novel animal models for the study of species-specific infectious diseases. Future applications of this new biotechnology include a variety of practical applications, such as

- Establishment of novel bioengineered animal models for the study of specific human pathogens

- Improved surgical procedures

- Novel site-specific delivery systems for cancer chemotherapy and gene therapy

CELL – TISSUE ELECTROFUSION TECHNOLOGY

Cell – Tissue Electrofusion Process

The process of cell–tissue electrofusion can be described in three steps (Fig. 6-1). The first step of the CTE process is to force fusion partners into close juxtaposition. This involves forcing the individual cells into contact with the top layer of cells of the tissue. Contact must be achieved in an electrically conductive media and between two electrodes. After contact has been achieved and while it is being maintained, cell–tissue fusion is induced by delivering one or more DC pulses to the cells and tissue. DC pulse(s) subject the cells and tissue to an electrical field. Cells exposed to electrical fields show an increase in transmembrane potential. When the transmembrane potential reaches a critical level the membranes are dielectrically broken down or porated (Bates et al., 1987; Zimmermann, 1982). If the membranes of two juxtaposed cells porate or break down in a region of mutual cell–tissue contact, then a lumen will be formed between the two cells. After pulse delivery, the third step occurs naturally to complete the fusion process. Normal membrane fluidity allows the fused partners to anneal into one cell.

In Vitro Cell – Tissue Electrofusion

Cell–tissue electrofusion can be accomplished in vitro with commercially available cell fusion instruments. For the *N. gonorrhoeae* model system

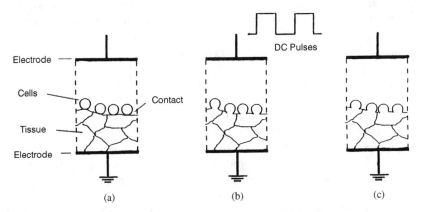

Figure 6-1. Three steps involved in the CTE process. (a) Step 1, force contact; (b) step 2, induce fusion via DC pulses; (c) step 3, cells in correct alignment anneal to form hybrid.

developed in our laboratory, specially designed chambers and electrodes were developed to facilitate the fusion of individual cells to intact tissue. The new design used mechanical force to juxtapose individual cells with intact tissue.

The in vitro chamber designed for the electrofusion of mechanically aligned cells consisted of six wells. Each insulated well has a flat, round 10-mm diameter platinum electrode in the bottom. The chamber could be attached to a water bath to maintain the desired thermal environment. The second electrode was a collection of six individual electrode heads of various shapes and sizes; each head could be fixed into a common insulated handle. The appropriate tip was selected to ensure maximum cell–tissue contact during the fusion process. The electrode tip shapes included a concave design that was machined to reflect the curvature of the rabbit cornea as well as 2 and 6 mm diameter polished titanium disks.

Procedure

The procedure using this CTE equipment is as follows. Briefly, a 10-mm corneal button was excised and placed in one of the insulated wells. A suspension of human lymphoma cells (U937 or HL60) was placed in contact with the excised tissue. The electrode contact with the cell–tissue system was controlled via a micromanipulator. Mechanical pressure was exerted through the electrode until the desired force was reached (600–700 g/cm^2). The force could be estimated by placing the entire apparatus on a torsion beam balance (or similar scale). The pressure was maintained for 3 min, then one-to-three 20-μsec, 0.20–2.0-kV/cm square pulses were delivered 1 sec apart to initiate the fusion process. The disk electrode was raised and the cornea removed from well. A vigorous wash and fixation in 2.5% glutaraldehyde followed. After fixation, the fusion results were examined by scanning electron microscopy (SEM).

Results

The above protocol was used to demonstrate the fusion of the two lymphoma cell lines to the rabbit corneal epithelium. The U937 cells were clearly visible on the surface of the rabbit corneal epithelium (Fig. 6-2) after the DC pulses were delivered. The U937 cells were easily distinguishable from the rabbit epithelial cells because of their spherical shape and smooth membranes that lack microvilli. In contrast, rabbit corneal epithelial cells were larger, were flat, and had centrally located nuclei and rough surfaces due to the presence of microvilli. Although the human cells appeared to be lying on the surface of the epithelium (Fig. 6-2a), when

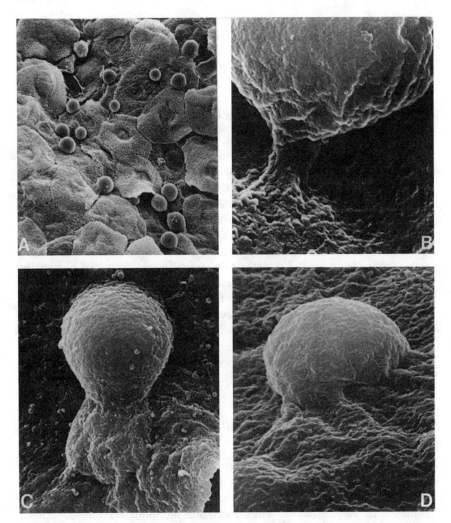

Figure 6-2. SEM of human U937 cells fused to the surface of rabbit corneal epithelium. (a) Corneal surface showing a representative distribution of 14 smooth spherically shaped human cells electrofused to the rabbit epithelium ($\times$1300); (b) plasma membrane "stalk" that forms very shortly after electrofusion formed between the upper smoother human cell and the lower microvilli covered rabbit cell surface ($\times$15,000 X); (c) and (d) illustrate coalescing human–rabbit plasma membranes of somatic cell hybrids beyond the initial stages of their formation in the intact corneal epithelium ($\times$6000; $\times$8600). [From Grasso et al., (1989); reproduced with permission.]

observed under higher magnification and at different angles (Fig. 6-2b–6-2d), several fusion points between the two cells were clearly discernable. At a different orientation, the two membranes appeared to be flowing together. From these observations, it was apparent that individual cells could be fused to intact tissue.

Transfer of Membrane Surface Components

The creation of model systems for the in vitro and in vivo study of species-specific infectious diseases will lead to a better understanding of the pathogenic mechanisms of a variety of receptor-mediated processes. CTE can be used to facilitate the transfer of distinctive features from a susceptible animal or human to a resistant animal. These animals would possess biological properties that would differ fundamentally from those displayed naturally by the unaltered animal species.

The first step in obtaining a model system is the establishment of a method that allows for the transfer of membrane surface components from susceptible cells to intact tissue of laboratory animals. CTE is a process that can be used to perform this transfer and establish novel model systems that can be used to study obligate human pathogens. The details of the transfer procedure and the usefulness of the system could be tested and characterized in vitro before attempting to establish an animal model. Therefore, an important consideration was to establish that transferred membrane components were present and functional following the fusion procedure.

Transfer of Neisseria gonorrhoeae Attachment Receptors

N. gonorrhoeae strain Pgh 3-2 was used to determine if gonococcal receptors present on human cells were present on the surface of newly formed human–rabbit somatic cell hybrids. Control experiments were performed, which included fusing nonhuman cells as well as corneas not subjected to electrofusion. After electrofusing human HL60 or human U937 lymphoma cells, rabbit skin cells, or Vero monkey kidney cells directly to the rabbit tissue, the corneas were used in qualitative gonococcal adherence assays and then examined by SEM. The gonococci attached only to those corneas that were fused with human cells. It is evident from these observations that the administration of an electrical field alone is not sufficient to modify the rabbit corneal epithelium to allow the attachment of this obligate human pathogen to the rabbit tissue. In addition, incorporation of nonhuman cells into the intact tissue did not allow attachment of the organism. Attachment was observed only when human cells were electrofused to the rabbit corneas (Fig. 6-3).

Figure 6-3. Attachment of *N. gonorrhoeae* to rabbit cornea. Receptor-mediated attachment of *N. gonorrhoeae* Pgh 3-2 to somatic human–rabbit cell hybrids formed within rabbit corneal epithelium by electrofusion. After human HL60 cells were electrofused to corneal epithelium, the corneas were used in gonococcal adherence assays and examined by SEM (×2000). [From Heller and Grasso (1990); reproduced with permission.]

Transfer of Human HLA Class I Antigens

The transfer of membrane surface components was examined further by determining if additional human specific markers could be detected on the rabbit corneal epithelium following CTE. It was important to ascertain if the transfer of the gonococcal receptors was an isolated fusion event or if other markers were also present following the fusion process. This series of experiments was performed by fusing human ME180 cervical carcinoma cells to rabbit corneas.

The ME-180 cells were first examined for the presence of the HLA class I antigens. The cells were examined by indirect immunofluorescence. ME-180 cells were exposed to mouse antihuman HLA class I (ABC antigens) antibodies followed by incubation with fluoresceinated goat antimouse IgG antibodies then analyzed with a FACScan flow cytometer. Greater than 95% of the cells were positive for the class I antigens in comparison with control cells treated only with fluoresceinated goat antimouse IgG antibodies.

The presence of HLA antigens on rabbit corneal epithelium after electrofusion with human ME-180 cells was determined. The corneas were incubated at 37°C for 30 min following the initiation of the fusion process. The histologically modified rabbit corneas were treated with mouse antihuman HLA class I antibodies and goat antimouse IgG antibodies conjugated to 0.5-μm microspheres. Corneas were exposed to each antibody for 30 min. Unfused corneas treated with both antibodies and fused corneas only exposed to the goat antimouse antibody were included as controls. All corneas were fixed in 2.5% glutaraldehyde and examined by SEM. Microspheres were found only on corneas containing fused human cells that had been treated with both antibodies (Fig. 6-4). Thus, human HLA class I markers present on ME-180 cells before fusion were present on the newly formed human–rabbit somatic cell hybrids.

In Vivo Cell – Tissue Electrofusion

Human HL60 cells were electrofused to the corneas of anesthetized rabbits to determine if cell–tissue electrofusion could be performed in vivo and to determine if it could eventually be used to develop an animal model for the study of *N. gonorrhoeae*. In addition, it was vital to ascertain if the CTE procedure produced an inflammatory response in the rabbits. It was also necessary to examine the rabbits for any long-term ill effects.

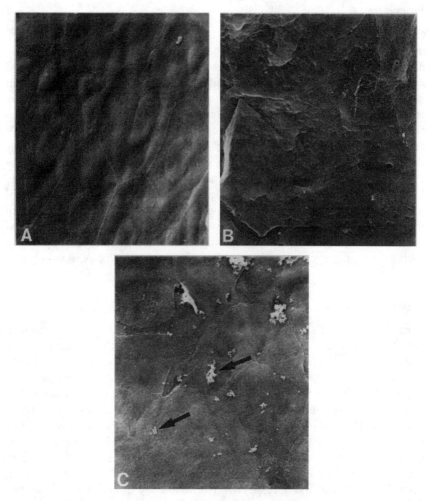

Figure 6-4. Human HLA class I antigens on rabbit corneas after fusion with human ME-180 cells. (a) SEM photomicrograph of unfused rabbit cornea treated with mouse antihuman HLA class I antibody and goat antimouse IgG conjugated to 0.5-μm microspheres ($\times$1800 X). (b) SEM photomicrograph of rabbit cornea after fusion with ME-180 cells and treated only with goat antimouse IgG conjugated to 0.5-μm microspheres ($\times$1800 X). Cornea was fixed with 2.5% glutaraldehyde 90 min after the initiation of the fusion process. (c) SEM photomicrograph of rabbit cornea after fusion with ME-180 cells and treated with both antibodies. Arrows point to microspheres indicating the presence of human HLA class I antigens on rabbit corneal surface. Cornea was fixed with 2.5% glutaraldehyde 90 min after the initiation of the fusion process ($\times$1800).

Procedure

Complete descriptions of the in situ and in vivo procedures are described in Grasso et al. (1989) and Heller and Grasso (1990). Briefly, human and nonhuman cells were washed by centrifugation in phosphate buffered saline and layered onto Millipore filters by centrifugation. Supernatant fluids were discarded, and the filters containing about 10^7 cells were removed with forceps and placed cell side down on the PBS-washed corneal surfaces of rabbits. Mechanical pressure plus three 20-μsec, 20-V square pulses with a duty cycle of 1 pulse/sec were applied simultaneously through a concave titanium electrode shaped to the curvature of the rabbit eye. For initial in situ and in vivo fusion studies, providing a field value (kilovolts per centimeter) was not practical because the ground potential was referenced to the rabbit's anatomy. Following the administration of the electrical pulses, all eyes were washed with PBS to remove unfused cells. Thirty minutes after electrofusing human HL60 cells directly to the rabbit tissue, the eyes were exposed to an infective dose of N. gonorrhoeae.

Results

Human HL60 cells were electrofused to the corneas of anesthetized rabbits to determine if CTE could be performed without harm to the animals. The electrofusion procedure used was the same as described above. A series of four experiments was set up, each experiment used four rabbits (Heller and Grasso, 1990). Briefly, rabbit No. 1 was set up as the electrofusion and human cell control. Fusion pulses were administered to the left eye in the absence of cells, while the right eye was exposed to human cells in the absence of fusion pulses. In all experiments, there were no observable signs of ocular inflammation in these eyes after exposure to the pathogen. Rabbit, murine, and human cells were electrofused to both eyes of rabbits 2, 3, and 4, respectively. The left eye of each animal was not exposed to gonococci and served as the inflammation control. There were no visible signs of inflammation detected in the left eyes of rabbits 2, 3, and 4. When the right eyes of rabbits 2 and 3 bearing electrofused nonhuman cells were exposed to the bacteria, neither inflammatory reactions nor ocular lesions developed. In striking contrast, a purulent keratoconjunctivitis developed in the right eye of rabbit 4 bearing electrofused human cells. Clinical symptoms of these ocular lesions appeared approximately 2 hr after infection and peaked after 8 and 12 hr (Fig. 6-5). Inflammation was completely resolved after 24 hr, and no additional inflammation was observed during an extended seven day observation period in any of the rabbits. This experimental design was repeated in each

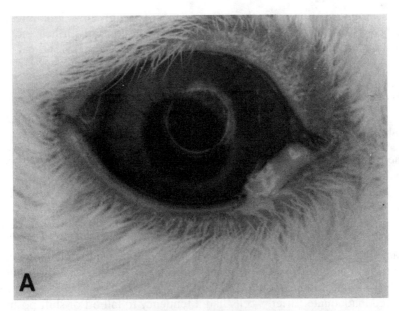

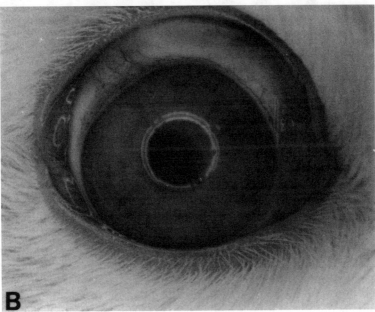

Figure 6-5. Gonococcal conjunctivitis in rabbit after transfer of gonococcal receptors by CTE. (a) Photograph of a mild gonococcal lesion that occurred in the right eye of Rabbit No. 4 only 4 hr after initiating the bacterial infection. Human–rabbit somatic cell hybrids were formed 30 min before the addition of the bacteria by electrofusing human HL60 cells to the rabbit corneal epithelium. (b) Photograph of the right eye of Rabbit No. 3 that had been similarly infected 4 hr earlier. Murine-rabbit somatic cell hybrids were formed 30 min before the addition of the bacteria by electrofusing murine WEHI-3 cells to the rabbit corneal epithelium. [From Heller and Grasso (1990); reproduced with permission.]

of the four experiments with similar results. The only time inflammation observed was if the rabbit eye had human cells fused to its surface and had been exposed to *N. gonorrhoeae*.

QUANTITATIVE CELL-TISSUE ELECTROFUSION

Cell Placement

The results from the initial cell–tissue electrofusion experiments, although qualitative, revealed an apparent variation in the number of cells fused to the tissue. This was due, in part, to the number of individual cells being placed in contact with the tissue before electrofusion. Control of the distribution and number of cells placed in contact with the tissue to be fused is an important consideration in minimizing the variability of CTE. Placing a suspension of cells in a defined volume onto the tissue and allowing the cells to settle by gravity did not yield an evenly distributed layer of cells on the tissue. Thus, Millipore filters were used, but they proved inadequate. Briefly, cells had a tendency to fall off or shift position when the filter was turned upside down to be placed onto the tissue. In addition, quantitation of the number of cells on these filters was hampered by the fact that fluorescently stained cells stuck to the filters and would not come off.

To overcome these problems, Zeta–Probe (ZP) blotting membranes, normally used in molecular biology (Church and Gilbert, 1984; Reed and Mann, 1985), were selected because they contain a positive electrostatic charge on their surface that has the ability to hold negatively charged cells in place during the fusion procedure. Cells placed on the ZP membranes can be quantitated either by staining cells with the vital fluorescent dye hydroethidine (HE; Bucana et al., 1986; Gallop et al. 1984) and measuring in a spectrofluorometer or by direct count with a hemocytometer. Cells were placed onto ZP disks as described in Heller and Grasso (1991). Briefly, ZP disks were placed in the wells of 48-well tissue culture plates. After 3×10^6 cells in 0.5-mL volume were added to the wells, the plates were centrifuged for 5 min at 180 g. The cells remained loosely adhered to the disks because of the electrostatic attraction between the negatively charged cells and positively charged disk. This attraction was sufficient to retain the cell during the removal of the disk from the plates and placement onto the surface of the tissue. Furthermore, the cells remain adhered to the disk at ambient temperatures for several hours. When the disks were removed from the plate and inverted to be placed on the tissue, none of the cells dropped from the disk but remained electrostatically adhered. Last, cells were not retained on the disk after DC electrofusion pulses were applied. There were no differences in any of the characteristics of non-HE-stained cells when subjected to these procedures. This was critically important to establish because quantitation was based upon HE-stained cells and the subsequent in vivo experiments were planned to be carried out with non-HE-stained cells.

Quantitation

To determine the number of cells fused quantitatively, a spectrofluorometric assay, sensitive enough to detect as little as 6×10^2, cells was developed. The assay was used to evaluate the reproducibility of the fusion process quantitatively. In addition, the assay would allow the initiation of examining important fusion parameters for eventual optimization of the process.

Procedure

The fluorometric assay for the quantitation of electrofusion (FAQE) was developed to allow a better characterization of this new technology (Heller, 1992). The basic approach that was adopted is outlined in the following steps. Specifically, (1) Cells in suspension were stained with hydroethidine, a vital fluorescent dye (Polysciences, Inc., Warrington, PA); (2) stained cells were washed by centrifugation to remove extracellular dye; (3) 10-mm-diameter corneal buttons were punched out; (4) stained cells layered on ZP disks were electrofused to the excised tissue (Gallop et al., 1984) (after placing the disks cell side down onto the surface of the epithelium, cell–tissue electrofusion was performed); (5) unfused stained cells were removed by washing the tissue; (6) stained fused cells on the corneal buttons were lysed by exposure to 0.2% (w/v) sodium dodecyl sulfate (SDS), which released the dye; (7) the remaining corneal tissue was removed and fluorescence intensities in the supernatants were measured with a spectrofluorometer; (8) the number of cells that had been electrofused was estimated from standard curves; and (9) these curves were generated from fluorescence measured in dilutions of lysed cells prepared from the same stained cell preparation that was used for electrofusion.

Results

The appropriate standard curves were used to estimate the numbers of HE-stained HL60 cells that had been electrofused to rabbit corneas in five separate experiments. Each experiment had nine corneas divided into three control groups and three corneas in the experimental group. The average number of fused cells per cornea was approximately 5000. In each of the five experiments, the standard deviation between the three experimental corneas was below 10%. In addition, the mean of these five experiments showed that all 15 experimental corneas had the same number of fused cells $+/- 10\%$ or less. Therefore, a reproducible number of cells could be electrofused to corneal epithelium. This spectrofluorometric assay appears quite acceptable for quantitating the process of cell–tissue electrofusion. These results demonstrate two main points; namely, this assay could now be used to analyze various fusion parameters quantitatively and several corneas could be prepared each with approximately the same number of cells fused to their surface.

Examination of Mechanical Force Necessary for CTE

CTE procedures use mechanical force to juxtapose cells before initiation of the fusion process. FAQE was used to examine the effect of various amounts of mechanical force on the number of cells fused (Heller, 1993). The amount of force applied to the system was measured in grams per electrode diameter by placing the fusion chamber on a torsion balance. ZP disks containing HL60 cells were placed cell side down onto rabbit corneas. Using a 2-mm-diameter electrode, between 25 and 400 g of force was applied to the cornea through the ZP disk. A single 20-μsec, 0.6-kV/cm square wave pulse was administered. The corneas were washed, fixed in 2.5% (v/v) glutaraldehyde, and examined by SEM. The results of these experiments indicate that a threshold is reached at 100 g (Table 6-1). The best results of fused cells were obtained at this level. Although a higher

TABLE 6-1. Effect of Mechanical Pressure on Cell-Tissue Electrofusion

Mechanical Pressure (g)	# Cells Fused (s.d.)
25	877
	(547)
100	4,693
	(250)
200	5,009
	(220)
400	6,649[a]
	(1,138)

Experiments were performed in triplicate.
[a] Damage to corneal tissue. [From Heller (1993); reproduced with permission.]

yield was obtained at 400 g, there was also a large amount of tissue damage. These results established a set of standard parameters to obtain reproducible CTE as follows: Cells on a ZP disk are placed on the tissue and a single 20-μsec, 0.6-kV/cm square wave pulse is administered through a flat, 2-mm-diameter circular electrode while applying 100 g of force (Heller and Gilbert, 1992).

TIME COURSE OF INCORPORATION

Results from the quantitative assay had shown the electromechanical parameters that would yield a reproducible number of cells fused onto intact tissue. These results allowed the use of multiple samples of corneas that were fixed, in 2.5% glutaraldehyde, at different times following the application of DC pulses. Four corneas were used for this set of experiments. Three corneas had cells fused to their surface at two spots with a 3-min interval between fusions. The three corneas were incubated at 37°C in DMEM without phenol red following fusion as follows: cornea No. 1, 0 min; cornea No. 2, 6 min; and cornea No. 3, 12 min. The fourth cornea served as the control and did not receive DC pulses but was incubated at 37°C in DMEM without phenol red for 12 min. From this experiment it was possible to observe the fusion process at 3-min time intervals from time 0 (negative control) to 18 min. SEM photomicrographs examining the incorporation of individual HL60 cells at 3, 6, 9, and 12 min following the application of DC pulses clearly show the individual cells coalescing into the epithelial cells (Fig. 6-6). These results reveal that by 15–20 min the individual cells are no longer clearly distinguishable from the corneal tissue and are completely incorporated (Heller, 1993).

SUMMARY

Cell–tissue electrofusion is a relatively new area of research that has evolved from an increasing number of investigations in the field of electrical manipulation of cells. Several important facts have already been established for CTE. Individual cells can be electrofused to intact tissue both in vitro and in vivo. Human gonococcal membrane receptors are present and remain functional after incorporation into human–rabbit somatic cell hybrids. Other human membrane components, HLA antigens, have also been shown to be transferred by the process of CTE. Individual cells are completely incorporated into intact tissue within 20 min. The number of cells fused can be quantitated fluorometrically. In addition, in vivo electrofusion can be performed without causing lethality or ocular inflammation to anesthetized animals. The CTE work presented also

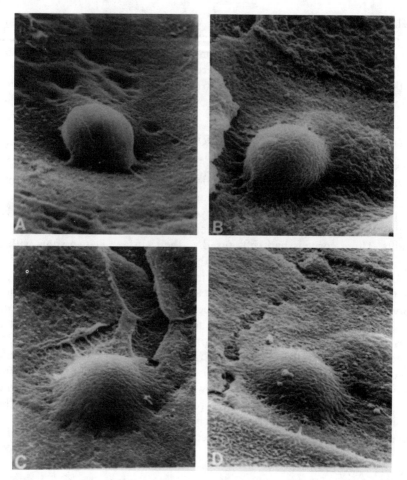

Figure 6-6. SEM photomicrograph showing time course of incorporation. Illustration of the various stages of incorporation of individual HL60 cells into intact rabbit corneal epithelium. (a) 3-min postfusion ($\times 4400$); (b) 6 min ($\times 4800$); (c) 9 min ($\times 4800$); (d) 12 min ($\times 4800$). [From Heller (1993); reproduced with permission.]

demonstrated that an obligate human pathogen can produce a purulent inflammatory response mediated by transferred human microbial attachment receptors in a common laboratory animal.

Demonstration that other human membrane components can be transferred by CTE and be present on a somatic cell hybrid is important for establishing the applicability of CTE. The ability to develop animal models for other species-specific diseases is dependent on demonstrating that other human membrane components can be transferred. For example, it was shown that CTE-treated rabbit corneas contained human specific HLA antigens after the fusion of human cells to their surface. To establish other models of infectivity, the correct cell type could be selected then fused to the desired animal or tissue. For example, to establish a model for HIV (AIDS) infectivity, H9 cells that contain CD4 on their surface could be fused to a selected tissue site to allow the binding of this pathogen.

Developing model systems for studying infectious diseases is the first of several applications for which CTE can be used. Two key factors in the development of these additional applications are the ability to optimize electrofusion procedures and to better understand the fusion process. Therefore, an ability to quantitate the number of cells fused using FAQE is an important part in the future development of CTE. Quantitation has already established the parameters necessary for obtaining reproducible fusion. Current work is focusing on examining many parameters important to the CTE process and finding combination(s) that will yield optimal CTE. It is envisioned that several new applications will emerge from this technology in the near future.

References

Abidor, I. G., and Sowers, A. E. (1992). Kinetics and mechanism of cell membrane electrofusion. Biophys. J. **61**: 1557–1569.

Bates, G., Saunders, J., and Sowers, A. E. (1987). Electrofusion principles and applications. Pages 367–395. In Cell Fusion. Sowers, A. E., ed. Plenum Press, New York.

Bucana, C., Saiki, I., and Nayar, R. (1986). Uptake and accumulation of the vital dye hydroethidine in neoplastic cells. J. Histochem. Cytochem. **34**: 1109–1115.

Chernomordik, L. V., Sukharev, S. I., Popov, S. V., Pastushenko, V. F., Sokirko, A. V., Abidor, I. G., and Chizmadzhev, Y. A. (1987). The electrical breakdown of cell and lipid membranes: The similarities of phenomenologies. Biochim. Biophys. Acta **902**: 360–373.

Church, G. M., and Gilbert, W. (1984). Genomic sequencing. Proc. Natl. Acad. Sci. USA **81**: 1991–1995.

Foung, S. H. K., and Perkins, S. (1989). Electric field-induced cell fusion and human monoclonal antibodies. J. Immunol. Meth. **116**: 117–122.

Frederik, P. M., Stuart, M. C. A., and Verkleij, A. J. (1989). Intermediary structures during membrane fusion as observed by cryo-electron microscopy. Biochim. Biophys. Acta **979**: 275–278.

Gallop, P. M., Paz, M. A., Henson, E., and Latt, S. A. (1984). Dynamic approaches to the delivery of reporter reagents into living cells. BioTechniques **2**: 32–36.

Glassy, M. (1988). Creating hybridomas by electrofusion. Nature **333**: 579–580.

Grasso, R. J., Heller, R., Cooley, J. C., and Haller, E. M. (1989). Electrofusion of individual animal cells directly to intact corneal epithelial tissue. Biochim. Biophys. Acta **980**: 9–14.

Heller, R. (1992). Spectrofluorometric assay for the quantitation of cell-tissue electrofusion. Anal. Biochem. **202**: 286–292.

Heller, R. (1993). Incorporation of individual cells into intact tissue by electrofusion. Pages 115–118. In Electricity and Magnetism in Biology and Medicine. Blank, M., ed. San Francisco.

Heller, R., and Gilbert, R. (1992). Development of cell-tissue electrofusion for biological applications. Pages 393–410. In Guide to Electroporation and Electrofusion. Chang, D. C., Chassy, B. M., Saunders, J. A., Sowers, A. E., eds. Academic Press, New York.

Heller, R., and Grasso, R. J. (1990). Transfer of human membrane surface components by incorporating human cells into intact animal tissue by cell-tissue electrofusion. Biochim. Biophys. Acta **1024**: 185–188.

Heller, R., and Grasso, R. J. (1991). Reproducible layering of tissue culture cells onto electrostatically charged membranes. J. Tiss. Cult. Meth. **13**: 25–30.

Hewish, D. R., and Werkmeister, J. A. (1989). The use of an electroporation apparatus for the production of murine hybridomas. J. Immunol. Meth. **120**: 285–289.

Kwekkeboom, J., de Groot, C., and Tager, J. M. (1992). Efficient electric field-induced generation of hybridomas from human B lymphocytes without prior activation in vitro. Human Antibod. Hybrid. **3**: 48–53.

Lo, M. M. S., and Tsong, T. Y. (1989). Producing monoclonal antibodies by electrofusion. Pages 259–270. In Electroporation and Electrofusion in Cell Biology. Neumann, A. E., Sowers, A. E., and Jordan, C. A., eds. Plenum Press, New York.

Lo, M. M. S., Tsong, T. Y., Conrad, M. K., Strittmatter, S. M., Hester, L. D., and Snyder, S. H. (1984). Monoclonal antibody production by receptor-mediated electrically induced cell fusion. Nature **310**: 792–794.

Neumann, E., Gerisch, G., and Opatz, K. (1980). Cell fusion induced by high electric impulses applied to *Dictyostelium*. Naturwissenschaffen **67**: 414–415.

Oberleithner, H. (1991). Epithelial cell fusion: New tool for cellular and molecular physiology. NIPS **6**: 181–184.

Ozawa, K., Hosoi, T., Tsao, C. J., Urabe, A., Uchida, T., and Takaku, F. (1985). Microinjection of macromolecules into leukemic cells by cell fusion technique: Search for intracellular growth-suppressive factors. Biochem. Biophys. Res. Commun. **130**: 257–263.

Prudovsky, I. A., and Tsong, T. Y. (1991). Fusion of fibroblasts with differentiated and nondifferentiated leukemia cells resulting in blockage of DNA synthesis. Dev. Biolog. **144**(2): 232–239.

Reed, K. C., and Mann, D. A. (1985). Rapid transfer of DNA from agarose gels to nylon membranes. Nucleic Acids Res. **13**: 7207–7221.

Richter, H.-P., Scheurich, P., and Zimmermann, U. (1981). Electric field-induced fusion of sea urchin eggs. Dev. Growth. Diff. **23**: 479–486.

Senda, M., Takeda, J., Abe, S., and Nakamura, T. (1979). Induction of cell fusion of plant protoplasts by electrical stimulation. Plant Cell Physiol. **20**: 1441–1443.

Sixou, S., and Teissie, J. (1992). In vivo targeting of inflamed areas by electroloaded neutrophils. Biochem. Biophys. Res. Comm. **186**: 860–866.

Sowers, A. E. (1987). The long-lived fusogenic state induced in erythrocyte ghosts by electric pulses is not laterally mobile. Biophys. J. **52**: 1015–1020.

Sowers, A. E. (1988). Fusion events and nonfusion contents mixing events induced in erythrocyte ghosts by an electric pulse. Biophys. J. **54**: 619–626.

Sowers, A. E. (1989a). The study of membrane electrofusion and electroporation mechanisms. Pages 315–337. In Charge and Field Effects in Biosystems—II. Allen, M.J., Cleary, S.F., and Hawkridge, F.M., eds. Plenum Press, New York.

Sowers, A. E. (1989b). Electrofusion of dissimilar membrane fusion partners depends on additive contributions from each of two different membranes. Biochim. Biophys. Acta **985**: 339–342.

Sowers, A. E. (1990). Low concentrations of macromolecular solutes significantly affect electrofusion yield in erythrocyte ghosts. Biochim. Biophys. Acta **1025**: 247–251.

Weber, H., Forster, W., Berg, H., and Jacob, H.-E. (1981a). Parasexual hybridization of yeasts by electric field simulated fusion of protoplasts. Curr. Genet. **4**: 165–166.

Weber, H., Forster, W., Jacob, H.-E., and Berg, H. (1981b). Microbiological implications of electric fields. III. Stimulation of yeast protoplast fusion by electric field pulses. Allg. Mikrobiol. **21**: 555–562.

White, J., Blackman, M., Bill, J., Kappler, J., Marrack, P., Gold, D. P., and Born, W. (1989). Two better cell lines for making hybridomas expressing specific T cell receptors. J. Immunol. **143**: 1822–1825.

Zimmermann, U. (1982). Electric field-mediated fusion and related electrical phenomena. Biochim. Biophys. Acta **694**: 227–277.

CHAPTER 7

Electrofusion of Plant Protoplasts for Hybrid Plant Production

M. G. K. Jones

ABSTRACT

Electrofusion has been applied to plant protoplasts with the production of a range of fusion products. These include symmetric hybrids,
such as somatic hybrids between *Solanum brevidens* and potato,
asymmetric hybrids (e.g., of tomato or tobacco), and the production of
cybrids, in which only the organelles (chloroplasts, mitochondria) are
transferred. The control of the fusion processes that can be achieved
by electrofusion makes this approach preferable to chemical fusion
procedures, particularly where delicate protoplasts are involved.

INTRODUCTION

There has been considerable interest in applying protoplast fusion technology to plants because of the potentially useful applications for genetic
manipulation of crop, forest, and horticultural species and its use in
furthering our understanding of somatic cell genetics.

The major approaches of protoplast fusion technology are as follows:

1. The combination of complete genomes of two species (symmetric fusion). The parental species may be sexually incompatible, and this

139

strategy may be used to try to extend the gene pool available to plant breeders. In other cases the parental protoplasts may be sexually compatible, with applications in genetic complementation studies or reconstruction of allopolyploids.

2. The production of asymmetric (partial) hybrids, either as a result of chromosome loss after symmetric fusion or after treatments designed to fragment the donor genome before fusion to the recipient protoplast.

3. The transfer of organelles between genotypes or species to introduce only chloroplast- or mitochondrion-encoded traits. The applications here include transfer of herbicide resistance (chloroplasts) and cytoplasmic male sterility traits (mitochondria).

Fusion of plant protoplasts can be achieved successfully both by chemical and electrical procedures. In this chapter, the emphasis is on experiences gained from the application of electrofusion specifically to plant protoplasts, and more theoretical aspects are only touched on briefly.

METHODOLOGY

General

Although the early work on electrofusion by Senda et al. (1979) was on plant protoplasts, much of the developmental work has been carried out with cells from the animal kingdom. In general, the approach for plant protoplasts is to suspend the protoplasts in a solution of low ionic strength with an osmoticum (e.g., 0.4 M mannitol) to prevent the protoplasts from bursting. A dielectrophoretic AC collection field (e.g., 1 MHz) is applied to align them into chains; then the DC fusion pulse is applied (Tempelaar and Jones, 1985a, 1985b, 1985c). For animal cells, because the cells are smaller than plant protoplasts, the electrode separation used has tended to be smaller (e.g., 100–200 μm). In the majority of cases, however, those working with plant protoplasts have found that a wider electrode spacing, usually 1 mm [although wider electrode separations of up to 7 mm have been used (Watts and King, 1984)], has proved to be more convenient to use. There are various reasons for this, the most obvious being

- Ease of manipulations
- Ease of observation and quantification and
- That more protoplasts can be treated at once

ELECTROFUSION PARAMETERS OF PLANT PROTOPLASTS

A general approach to the production of somatic hybrid plants by protoplast fusion is outlined in Figure 7-1.

Several factors affect the efficiency of electrofusion of plant protoplasts. The majority of these factors must be considered whenever plant protoplasts are isolated and cultured; others are specific considerations for electrofusion. They include the following:

- Species of source plant
- Growth and age of the plant or culture

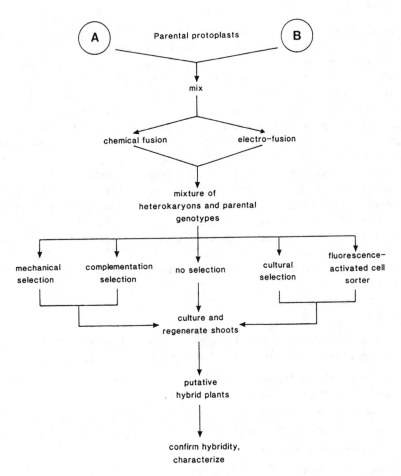

Figure 7-1. Overall procedures involved in fusion and regeneration of somatic hybrid plants following fusion of parental protoplasts.

- Source tissues (e.g., leaf, cell suspension, hypocotyl)
- Protoplast isolation procedure (enzymes, medium, osmolarity)
- Purity and viability of the protoplasts
- Fusion medium (osmolarity, additives e.g., Ca^{2+} ions, spermine)
- Density of protoplasts
- Membrane properties of the protoplasts
- Geometry of the fusion chamber
- Electrofusion field parameters
- Diameter of the protoplasts

One attraction of electrofusion is that the greater degree of control of the process when compared with chemical fusion methods allows some of these variables to be analyzed and understood in more detail. Analytical studies can be carried out using a simple electrode chamber consisting of two wires stretched across a glass slide, 1 mm apart (Tempelaar and Jones, 1985a). Following application of the dielectrical collection field and fusion pulse(s), the chains of protoplasts can be maintained in place with an AC field of about 30% of that used for alignment before fusion. This allows ready quantification of fusion events, and the following parameters can be measured: percentage of total aligned protoplasts fused, number of 1 : 1 (binary) fusions and multiple fusions, effects of varying fusion pulse voltage and duration, effects of additives and ionic composition of fusion medium, effects on fusion of different protoplast isolation procedures, effects of varying the ratios of parental protoplasts (Tempelaar and Jones 1985a, 1985b, 1986c; Tempelaar et al. 1987; Fish et al. 1988c). This approach allows rapid optimization of electrofusion parameters, and examples of such analysis are given in Figures. 7-2 and 7-3.

From such studies, the following factors influence the fusion frequency obtained:

- Tissue source of the protoplasts; in general, leaf protoplasts fuse more readily than protoplasts from cell suspension cultures (Fig. 7-2).
- Protoplast diameter; in general larger protoplasts can be fused more easily than smaller protoplasts (Fig. 7-2).
- Length of the chains of aligned protoplasts influences fusion frequencies; longer chains give higher fusion frequencies than short chains (Fig. 7-3). Chain length can be influenced by
- Protoplast density
- Amplitude and length of time the alignment field is applied

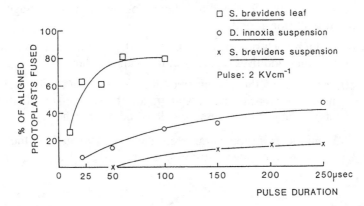

Figure 7-2. Pulse duration–fusion response curves for self-electrofusion of leaf and cell-suspension protoplasts. This shows that leaf protoplasts of *S. brevidens* are more fusogenic than suspension cell protoplasts of the same genotype and that the larger *Datura innoxia* protoplasts are more fusogenic than the smaller *S. brevidens* protoplasts.

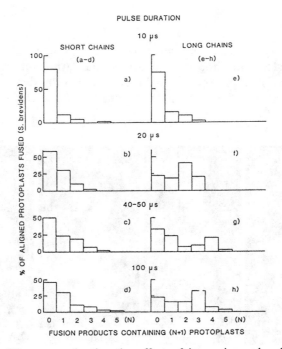

Figure 7-3. Histograms showing the effect of increasing pulse durations with short- and long-aligned protoplast chains on the number of 1 : 1 or multiple-fusion products produced (from Tempelaar and Jones, 1985a).

- Fusion pulse length and its voltage; longer pulses at higher voltages lead to more multifusion products.
- Fusion medium; inclusion of 1 mm Ca^{2+} in the medium improves fusion frequency; other additives (e.g., spermine) can also help.
- Osmotic pressure of the fusion medium; lower osmotic pressures give higher fusion frequencies but result in lower viability (or bursting of protoplasts) if the osmotic pressure is lowered too much.

Once optimum electrofusion parameters have been identified, these can be applied to whatever scale of fusion is desired (Fig. 7-4).

Scale

The scale of fusion experiments that have been studied for plant protoplasts has ranged from pairs of individual protoplasts (microfusion) to mass fusion (macrofusion).

Microfusion

One attraction of fusing selected pairs of protoplasts is that the need for postfusion selection procedures is overcome. An alternative problem, however, that plant protoplasts do not grow well at low densities unless the culture volume is reduced significantly, needs to be solved.

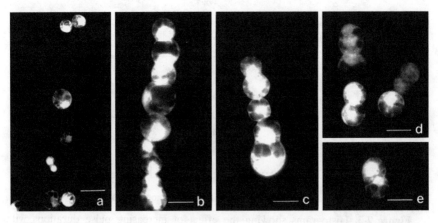

Figure 7-4. Wheat cell suspension protoplasts stained with fluorescein undergoing electrofusion.

In the initial report of Senda et al. (1979), microelectrodes were used to deliver fusion pulses to protoplasts that had aggregated spontaneously. This was followed by the work of Koop et al. (1983) who fused selected protoplasts in 100-nL droplets of 0.4M mannitol using platinum electrodes to deliver a dielectrical collection field, followed by a single DC fusion pulse. This approach was subsequently refined and automated, culminating in the elegant system reviewed in Schweiger et al. (1987). The refinements to the system were as follows:

- An improved microculture system with computer control
- Efficient procedures for selection, transfer, and handling of individual protoplasts or fusion products
- Electrofusion of defined protoplast pairs

The procedure is outlined in Figure 7-5.

The microculture system consists of mineral oil droplets (to prevent evaporation) into which aqueous droplets of medium of volumes 15–100

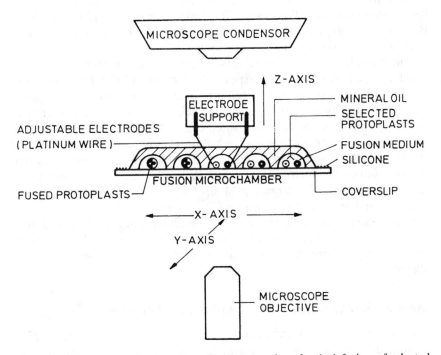

Figure 7-5. Automated microelectrofusion procedure for 1:1 fusion of selected protoplasts or cytoplasts in aqueous droplets under oil (redrawn from Schweiger et al., 1987).

nL are injected. One protoplast in 50 nL of medium is equivalent to 2×10^4 protoplasts per milliliter on mass culture. For electrofusion, two selected protoplasts are delivered into a droplet of low ionic strength medium, then the droplet is positioned under platinum wire electrodes, which are held at a fixed distance apart. The electrode support is lowered so that the electrodes enter the aqueous droplet. Following application of the AC collecting field to bring the protoplasts together, the DC fusion pulse induces fusion. The fusion product is then transferred to a droplet of full culture medium for longer term culture and regeneration. Fusion frequencies of 45–60% were reported for this approach (Schweiger et al. 1987), which has also been used to fuse subprotoplasts and enucleated cytoplasts to protoplasts and to fuse microinjected protoplasts.

Macrofusion

The alternative to the exacting approach of handling, fusing, and culturing selected pairs of protoplasts is mass fusion. In this approach, large num- bers of protoplasts are aligned and fused. This is usually accomplished in batch treatments (e.g., Watts and King, 1984; Tempelaar and Jones, 1985b; Bates et al., 1987; Fish et al., 1988a), although flow cell systems have also been developed (Hibi et al., 1988).

For example, using five parallel stainless-steel electrodes 1 mm apart, which can be inserted into a perspex well glued to a glass microscope slide, $0.5–1 \times 10^6$ protoplasts can be fused in a cycle lasting 60–90 sec (Fish et al., 1988a). The fusion products may be selected after this treatment; for example, using a fluorescence-activated cell sorter (FACS) or mechanically using a micropipette, or they may be cultured without selection and hybrid shoots identified once they regenerate. The high fusion frequencies that can be attained by electrofusion reduce the problems of identifying hybrid shoots and reduce the requirement for more elaborate selection proce- dures to be applied just after fusion (Fish et al., 1988a).

By combining knowledge of fusion characteristics in electrical fields with different ratios of parental protoplasts, in some cases it is also possible preferentially to produce heterofusion products (Tempelaar and Jones, 1985b).

Mass Fusion without Dielectrophoretic Alignment

It is also possible to electrofuse plant protoplasts in full culture medium without dielectrophoretic alignment, by allowing the protoplasts to sedi- ment on the bottom of the culture dish as a lawn, perhaps in the presence of polyethylene glycol (to promote close contact). A DC fusion pulse can then be applied to induce pore formation and fusion (Montané et al., 1987).

PRODUCTION OF HYBRID PLANTS BY ELECTROFUSION

Symmetrical Hybridization

The production of somatic hybrid plants following electrofusion of complete genomes of parental protoplasts has been achieved for a range of combinations of protoplasts. In the case of hybrids produced by fusion of protoplasts of sexually incompatible parents, novel hybrids have been produced. The approach has also been applied to the resynthesis of allopolyploid plants by fusion of closely related species and to complementation studies between mutant lines of the same species. Selected examples of these approaches, where electrofusion has been used to produce hybrids, are provided.

The somatic hybrids produced by electrofusion of the sexually incompatible South American wild *Solanum* species. *S. brevidens* and dihaploid potato (*S. tuberosum*), provide a well-characterized example of novel hybrids (Fig. 7-6). The interest in *S. brevidens* is that it is a source of resistance to the major virus diseases of potato (Potato Leaf Roll Virus, Potato Virus X, and Potato Virus Y) and some genotypes also confer

Figure 7-6. Left: dihaploid potato plant; right: diploid *S. tuberosum*; center: tetraploid somatic hybrid from electrofusion of potato + *S. brevidens* (from Fish et al., 1988a).

resistance to the soft rot bacterium *Erwinia carotova*. Somatic hybrids between *S. brevidens* and *S. tuberosum* have been produced both by chemical procedures (e.g., Barsby et al., 1984; Austin et al., 1985; Fish et al., 1987) and by electrofusion (Fish et al., 1988a). When a direct comparison between the two approaches was made, electrofusion was found to be the more efficient approach. On analysis of shoots regenerated (by isoenzyme analysis or DNA probes) without selection following mass fusion using a chemical procedure, 2.3% of regenerants were hybrids, whereas after electrofusion the figure was 12.6% (Fish et al. 1988a). The analysis of 60 somatic hybrids resulting from this work gives a broad picture of the characteristics of plants that can be produced following experimental fusion.

Cytology

Analysis of root tip chromosomes of *S. brevidens* (24 chromosomes) and *S. tuberosum* (dihaploid, 24 chromosomes) gave the following results:

	Percentage of Hybrid Plants
Tetraploid (24 + 24 chromosomes, products of binary fusions)	20.3%
Hexaploid (24 + 24 + 24 chromosomes, products of triple fusions)	14.1%
Aneuploid around the tetraploid level (43–56 chromosomes)	23.4%
Aneuploid around the hexaploid level (61–79 chromosomes)	32.8%
Aneuploid approaching the octaploid level (about 85 chromosomes)	9.4%

These results indicate that products of three protoplasts fusing together are viable and can be stable, but products from four protoplasts fusing are unstable. The aneuploids may be the result of either some incompatibility between the parental species or simply of somaclonal (i.e., tissue culture induced) variation. Plants regenerated from potato protoplasts without any fusion treatment can be 40% aneuploid (Fish and Karp, 1986).

Chloroplasts

By extraction of chloroplast DNA and separation of fragments produced by cutting it with specific restriction enzymes, it is relatively easy to find size fragment patterns characteristic of the parental genotypes. When somatic hybrids are analyzed for the origin of their chloroplasts, in the case of *S. brevidens* + *S. tuberosum*, 55% had chloroplasts from *S. brevidens* and 45% from *S. tuberosum*. There was no evidence of recombination, segregation appeared to be random, and there was no correlation with genome dosage in hexaploid hybrids. Thus, six classes of euploid somatic hybrids were obtained (Pehu et al., 1989):

Tetraploid Euploid Hybrids (48 chromosomes)

1. Nucleus: *S. brevidens* + *S. tuberosum*   Chloroplasts: *S. tuberosum*
2.   *S. brevidens* + *S. tuberosum*   *S. brevidens*

Hexaploid Euploid Hybrids (72 chromosomes)

3. Nucleus: 2 × *S. brevidens* + 1 × *S. tuberosum*   Chloroplasts: *S. tuberosum*
4.   2 × *S. brevidens* + 1 × *S. tuberosum*   *S. brevidens*
5.   1 × *S. brevidens* + 2 × *S. tuberosum*   *S. tuberosum*
6.   1 × *S. brevidens* + 2 × *S. tuberosum*   *S. brevidens*

RFLP Analysis of Nuclear Genome

A series of molecular probes (e.g., pGM01—patatin cDNA or SSd12—a random cDNA probe) can be used to differentiate between *S. brevidens* and *S. tuberosum* by Southern hybridization and to verify hybridity of these somatic hybrids (Pehu et al., 1989). Indeed, the intensity of the species-specific bands was used to analyze the parental genome dosages in the hexaploid hybrids of the previous section.

Virus Resistance

Analysis of the somatic hybrid plants for resistance to PLRV, PVY, and PVX from *S. brevidens* showed a range of resistances from high to low, with the low resistance correlating with aneuploidy or plants that resembled *S. tuberosum* more than *S. brevidens*. Study of virus resistance in hypoaneuploid hybrids (which had lost from one to three chromosomes) showed that loss of resistance to the three viruses occurred together, so the resistances are closely linked and may have a common basis. Further study of virus replication in *S. brevidens* protoplasts showed that the viruses could replicate equally well if introduced directly into protoplasts and suggests that the resistance mechanism may be an inhibition of cell-to-cell spread of the viruses (Pehu et al., 1990; Gibson et al., 1990; Valkonen et al., 1991).

Species-Specific Probes

As an aid to rapid identification of symmetric or asymmetric hybrid plants, a series of DNA probes was developed that hybridized specifically to DNA from *S. brevidens* or from *S. tuberosum* (Pehu et al., 1991). These can be used to do the following:

1. Verify hybridity
2. Determine parental genome dosage (in hexaploid hybrids)
3. Identify hybrid callus/tissues by application of a squash blot approach (Pehu et al., 1991)

Tomato / Potato RFLP Markers

The marker sequences developed for RFLP Mapping of tomato (Bonierbale et al., 1988) and potato (Gebhardt et al., 1989) can also be used to analyze hybrids in more detail (Williams et al., 1991; unpublished data) and are of particular use in the study/production of asymmetric hybrids as detailed in a later section.

Field Growth of Somatic Hybrids

The somatic hybrids between *S. brevidens* and *S. tuberosum* produced by electrofusion were grown in field trials in the United Kingdom from 1986 to 1990 (Fish et al., 1988b). They were classified as genetically manipulated organisms and were therefore subject to the same guidelines and controls as plants containing recombinant DNA (laid down by the UK Advisory Committee for Genetic Manipulation). The morphology of these somatic hybrids and their tuberization have been studied in detail (Fish et al., 1988b, Pehu et al., 1989, unpublished, Fig. 7-7). It is worth noting that similar somatic hybrids, produced by chemical fusion and grown in the field in Wisconsin, were not subject to these restrictions (Austin et al., 1986). The hybrid plants are essentially male infertile but can be used as female parents in backcrosses to *S. tuberosum* in order to make use of the agronomically useful genes from *S. brevidens*.

Asymmetric Hybridization

One of the disadvantages of interspecific or intergeneric somatic hybrids in which there is complete combination of the nuclear genomes is that there

Figure 7-7. Field-grown tubers of a somatic hybrid plant of *S. brevidens* and potato.

is transfer of unwanted genetic material as well as the desired characters. Work with *S. brevidens* illustrates this well. It is nontuber bearing, so tubers produced by somatic hybrids with potato tend to be elongated, with pronounced eyes and reduced dormancy. The production of asymmetric hybrids, in which only single chromosomes or subchromosomal fragments are transferred, would solve this problem without the need to isolate and manipulate gene sequences. This approach is well illustrated by the work of Bates (1990) on tobacco and Wijbrandi and Koornneef (1990) and Derks et al. (1991) for tomato.

The approach normally used to produce asymmetric hybrids is to irradiate donor protoplasts and then to fuse them with recipient protoplasts (donor–recipient fusions). Irradiation serves both to fragment the donor genome and to block subsequent division of the donor protoplasts. The work of Bates and co-workers with tobacco, who used electrofusion to create asymmetric hybrid plants, illustrates this approach (Bates et al.,

1987; Bates, 1990). In the first of these publications, protoplasts of kanamycin-resistant *Nicotiana plumbaginifolia* were irradiated and fused with protoplasts of *N. tabacum*. Selection on kanamycin-containing medium led to regeneration of plants resembling *N. tabacum* that retained the kanamycin resistance trait from *N. plumbaginifolia*. These hybrids were male sterile, but were backcrossed to *N. tabacum* and produced kanamycin-resistant progeny. Analysis using species-specific DNA probes showed that one to several *N. plumbaginifolia* chromosomes were present. A similar approach was applied to transfer useful disease-resistance characters from *N. repandra* and *N. glutinosa* (both with kanamycin resistance) to *N. tabacum*. Further characterization of the asymmetric *N. tabacum* (+) *N. repanda* hybrids showed that most plants had 56–62 chromosomes, although the number ranged from 41 to 90. Since tobacco has 48 chromosomes, if it is assumed that all chromosomes in hybrids over 48 originated from *N. repandra*, then these hybrids retain 4–88% of the donor chromosomes, with the majority at about 23%. These asymmetric hybrids were female fertile but male sterile, whereas similar asymmetric hybrids with *N. glutinosa* as the donor were both female and male fertile.

Studies on production of asymmetric hybrids, such as those cited above, have generally not resulted in recipient plants that only contain one chromosome or chromosomal fragment from the donor, which is the real goal for this work. One reason is that the extent of chromosome elimination from the donor does not always seem to be directly correlated with the dose of radiation given. If a selection system is used to obtain asymmetric hybrids, then it will only be of use if the selectable trait (e.g., kanamycin resistance) is linked to the desirable trait to be transferred. If a selectable trait is not used, then many asymmetric hybrids must be regenerated and tested for the desired character. Whatever the approach used, the smaller the genetic fragments transferred, the larger will be the number of hybrids that must be examined. Nevertheless, the research so far carried out indicates that it will be possible to transfer specific unselected traits by asymmetric hybridization (Bates, 1990).

Production of Cybrids

Maternal inheritance of cytoplasmic organelles usually occurs after sexual crosses between plants. As indicated in the studies on the somatic hybrids between *S. brevidens* and *S. tuberosum*, protoplast fusion combines both cytoplasmic genomes and the nuclear genomes. There are both theoretical and practical reasons for wanting to transfer chloroplast and mitochondrial traits into different nuclear backgrounds. For example, it is of interest in studying nuclear–organelle interactions and in the transfer of herbicide-resistant chloroplasts, or, in the case of oilseed rape (*Brassica napus*), the transfer of mitochondria from *Raphanus sativa* that confer cytoplasmic male sterility (cms) while maintaining the *B. napus* chloroplasts, so that cms lines can be used as female parents in the production of hybrid rape seed.

The transfer of organelles from a donor to a recipient protoplast by fusion can be an extension of asymmetric hybridization, in which the donor nuclear genome can be completely removed by strong irradiation, which does not destroy organelle genomes. It can also be achieved by fusion of enucleated protoplasts with recipient protoplasts. The latter approach has been used effectively by Spangenberg et al. (1991), who used the microelectrofusion procedures previously described (Schweiger et al., 1987) to fuse individual cytoplasts (enucleated protoplasts) with protoplasts. In this case, one parental *N. tabacum* had a kanamycin-resistant nuclear genome and streptomycin-resistant chloroplasts, whereas the other had a hygromycin-resistant nuclear genome and a cms mitochondrial marker. The latter line was used to produce enucleated cytoplasts that were electrofused with protoplasts of the former line. Of 30 regenerants analyzed in detail, 29 were cybrids of the expected nucleus (kanamycin resistant, hygromycin sensitive). Of these, about 20% had streptomycin-resistant chloroplasts and the remainder were streptomycin-sensitive cybrids. These results further support the general observation that chloroplasts sort after fusion so that only one type is found in regenerants (although chimeras occasionally occur). In contrast, analysis of mitochondrial DNA indicated that the mitochindria in regenerants could be that of either parent, additive, or novel as a result of recombination.

This work elegantly demonstrates that microelectrofusion can be used for the production of defined cybrids of higher plants.

SUMMARY

The examples given show that electrofusion of plant protoplasts and cytoplasts is being used widely to produce symmetric hybrids, asymmetric hybrids, and cybrids and that these approaches are being used to further our understanding of basic aspects of somatic cell genetics, to extend the combinations of genetic material available, and as an additional tool for plant breeders. The control over the electrofusion process when compared with chemical procedures has been a significant advance and has led, for example, to the development of elegant microfusion procedures and to higher fusion frequencies in mass fusions. The latter have reduced the need for more rigorous postfusion selection procedures, since 10% or more of the regenerants can be hybrid following macroelectrofusion. In addition, electrofusion can also be used to study membrane properties of plant protoplasts.

References

Austin, S., Baer, M. A., and Helgeson, J. P. (1985). Transfer of resistance to potato leaf roll virus from *Solanum brevidens* into *Solanum tuberosum* by somatic fusion. Plant Sci. **39**: 75–82.

Austin, S., Ehlenfeldt, M. K., Baer, M. A., and Helgeson, J. P. (1986). Somatic hybrids produced by protoplast fusion between *S. tuberosum* and *S. brevidens*: Phenotypic variation under field conditions. Theor. Appl. Genet. **71**: 628–690.

Barsby, T. L., Shepard, J. F., Kemple, R. J., and Wong, R. (1984). Somatic hybridization in the genus *Solanum*: *S. tuberosum* and *S. brevidens*. Plant Cell Rep. **3**: 165–167.

Bates, G. W. (1990). Transfer of tobacco mosaic virus resistance by asymmetric protoplast fusion. Pages 293–298. In Proc. VII Inter. Congress on Plant Tissue and Cell Culture, Amsterdam, The Netherlands, 24–29 June. Progress in Plant Cellular and Molecular Biology. Nijkamp, H. J. J., Van Der Plas., L. H. W., and Van Aartrijk, J., eds., Kluwer Academic Publishers, Dordrecht, The Netherlands.

Bates, G. W., Hasenkampf, C. A., Contolini, C. L., and Piastuch, W. C. (1987). Asymmetric hybridization in *Nicotiana* by fusion of irradiated protoplasts. Theor. Appl. Genet. **74**: 718–726.

Bonierbale, M. W., Plaisted, R. L., and Tanksley, S. D. (1988). RFLP maps based on a common set of clones reveal modes of chromosomal evolution in potato and tomato. Genetics **120**: 1095–1103.

Derks, F. H. M., Wijbrandi, J., Koornneef, M., and Colijn-Hooymans, C. M. (1991). Organelle analysis of symmetric and asymmetric hybrids between *Lycopersicon peruvianum* and *Lycopersicon esculentum*. Theor. Appl. Genet. **81**: 199–204.

Fish, N., and Karp, A. (1986). Improvements in regeneration from protoplasts of potato and studies on chromosome stability. Theor. Appl. Genet. **72**: 405–412.

Fish, N., Karp, A., and Jones, M. G. K. (1987). Improved isolation of dihaploid *S. tuberosum* protoplasts and the production of somatic hybrids between dihaploid *S. tuberosum* and *S. brevidens*. In Vitro Cell and Dev. Biol. **23**: 575–580.

Fish, N., Karp, A., and Jones, M. G. K. (1988a). Production of somatic hybrid plants of *Solanum* by electrofusion. Theor. Appl. Genet. **76**: 260–266.

Fish, N., Lindsey, K., and Jones, M. G. K. (1988c). Plant protoplast fusion. Pages 481–498. In Methods in Molecular Biology, Vol. 4. Walker, J.M., ed. Humana Press, Clifton, New Jersey.

Fish, N., Steele, S., and Jones, M. G. K. (1988b). Field characteristics of somatic hybrid plants of *S. tuberosum* and *S. brevidens*. Theor. Appl. Genet. **76**: 880–886.

Gebhardt, C., Blonendahl, C., Schactschable, U., Debener, T., Salamini, F., and Ritter, E. (1989). Identification of 2n breeding lines and 4n varieties of potato (*Solanum tuberosum* spp. *tuberosum*) with RFLP fingerprints. Theor. Appl. Genet. **78**: 16–22.

Gibson, R. W., Jones, M. G. K., and Fish, N. (1988). Resistance to potato leaf roll virus and potato virus Y in somatic hybrids between diphaloid *S. tuberosum* and *S. brevidens*. Theor. Appl. Genet. **76**: 113–117.

Gibson, R. W., Pehu, E., Wood, R. D., and Jones, M. G. K. (1990). Resistance to potato virus Y and potato virus X in *Solanum brevidens*. Ann. Appl. Biol. **116**: 151–156.

Hibi, T., Kano, H., Sugiura, M., Kazami, T., and Kimura, S. (1988). High-speed electrofusion and electrotransfection of plant protoplasts by a continuous flow electromanipulator. Plant Cell Rep. **7**: 153–157.

Jones, M. G. K. (1988). Electrofusion of plant protoplasts. Trends in Biotechnol. **6**: 153–158.

Jones, M. G. K., Dunckley, R., Steele, S., Karp, A., Gibson, R., Fish, N., Valkonen, J., Poutala, T., and Pehu, E. (1990). Transfer of resistance to PLRV, PVX and PVY from *S. brevidens* to potato by somatic hybridization: Characterisation and field evaluation. Pages 286–293 In Proc. VII Inter. Congress on Plant Tissue and Cell Culture, Amsterdam, The Netherlands, 24–29 June. Progress in Plant Cellular and Molecular Biology. Nijkamp, H. J. J., Van Der Plas, L. H. W., and Van Aartrijk, J., eds. Kluwer Academic Publishers, Dordrecht, The Netherlands.

Koop, H-U., Dirk, J., Wolff, D., and Schweiger, H. G. (1983). Somatic hybridization of two selected single cells. Cell Biol. Int. Rep. **7**: 1123–1128.

Montané, M. H., Ailbert, G., and Teisié, J. (1987). Genetic investigations of somatic hybrids between tobacco albino strains obtained by electrofusion. Studia Biophysica **119**: 89–92.

Pehu, E., Gibson, R. W., Jones, M. G. K., and Karp, A. (1990). Studies on the genetic basis of resistance to PLRV, PVX, and PVY in *Solanum brevidens* using somatic hybrids of *S. brevidens* and *S. tuberosum*. Plant Sci. **69**: 95–101.

Pehu, E., Karp, A., Moore, K., and Jones, M. G. K. (1989). Molecular, cytological and morphological characterisation of somatic hybrids of dihaploid *Solanum tuberosum* and diploid *S. brevidens*. Theor. Appl. Genet. **78**: 696–704.

Pehu, E., Thomas, M., Poutala, T., Karp, A., and Jones, M. G. K. (1991). Species-specific sequences in the genus *Solanum*: Identification, characterization and application to study somatic hybrids of *S. brevidens* and *S. tuberosum*. Theor. Appl. Genet. **80**: 693–698.

Schweiger, H. G., Dirk, J., Koop, H. U., Kranz, E., Spangenberg, G., and Wolff, G. (1987). Individual selection, culture and manipulation of higher plant cells. Theor. Appl. Genet. **73**: 769–783.

Senda, M., Takeda, J., Abe, S., and Nakamura, T. (1979). Induction of cell fusion of plant protoplasts by electric stimulation. Plant Cell Physiol. **20**: 1441–1443.

Spangenberg, G., Freydl, E., Osusky, M., Nagel, J., and Potrykus, I. (1991). Organelle transfer by microfusion of defined protoplast-cytoplast pairs. Theor. Appl. Genet. **81**: 477–486.

Tempelaar, M. J., Duyst, A., de Vlas, S. Y., Krol, G., Symonds, C., and Jones, M. G. K. (1987). Modulation and direction of the electrofusion response in plant protoplasts. Plant Sci. **48**: 99–105.

Tempelaar, M. J., and Jones, M. G. K. (1985a). Fusion characteristics of plant protoplasts in electric fields. Planta **165**: 205–216.

Tempelaar, M. J., and Jones, M. G. K. (1985b). Directed electrofusion between protoplasts with different responses in a mass fusion system. Plant Cell

Rep. **4**: 92–95.

Tempelaar, M. J., and Jones, M. G. K. (1985c). Analytical and preparative electro-fusion of plant protoplasts. Pages 347–351. In Oxford Surveys on Plant Molecular and Cell Biology. Vol. 2. Miflin, B. J., ed. Oxford University Press, Oxford.

Valkonen, J. P. T., Pehu, E., Jones, M. G. K., and Gibson, R. W. (1991). Resistance in *Solanum brevidens* to both potato virus Y and potato virus X may be associated with slow cell-to-cell spread. J. Gen. Virol. **72**: 231–236.

Watts, J. W., and King, J. M. (1984). A simple method for large scale electrofusions and culture of plant protoplasts. Biosci. Rep. **4**: 335–342.

Wijbrandi, J., and Koorneef, M. (1990). Partial genome transfer in interspecific tomato hybrids. Pages 280–285. In Proc. VII International Congress on Plant Tissue and Cell Culture, Amsterdam, The Netherlands, 24–29 June. Progress in Plant Cellular and Molecular Biology. Nijkamp, H. J. J., Van Der Plas, L. H. W., and Van Aartrijk, J., eds. Kluwer Academic Publishers, Dordrecht, The Netherlands.

CHAPTER 8

Gene Delivery by
Membrane Electroporation

E. Neumann

ABSTRACT

Gene delivery by electroporation can be described in physical-chemical terms of an electroporation-resealing hysteresis, with unidirectional state transitions coupled to electrodiffusive migration of DNA through cell wall structures and electroporated plasma membranes. Deeper insight into electroporation phenomena such as electrotransfection, electrofusion, and electroinsertion is gained by the inspection of electrosensitivity or recovery curves of cell populations as well as by the analysis of pulse intensity–duration relationships. A theoretical framework is outlined for an adequate comparison of data obtained with different pulse shapes. The results of the physical-chemical analysis not only indicate possible mechanisms but also are instrumental for goal-directed optimization strategies for practical applications of electroporation techniques.

INTRODUCTION

Membrane electroporation is a well-established method for the electrical manipulation of cells and organelles of all types of microorganisms and tissues (Sowers, 1987; Neumann et al., 1989; Chang et al., 1991). The electrical parameters of the impulses to be applied in electroporation techniques are adjusted according to cell size and cell type. Of particular importance for the parameter choice is the presence of cell walls or extracellular and intracellular matrix structures. Therefore, the electrical field strengths and pulse durations cover the side range of about 0.2–30 kV cm^{-1} and of about 0.01–30 msec, respectively.

Under the correct conditions of pulse characteristics and medium composition, electroporation causes transient and reversible (i.e., nondestructive) permeability changes in patches of cell surface membranes (Neumann and Rosenheck, 1972, 1973; Zimmermann et al., 1973, 1974). This electropermeabilization can be used to induce transmembrane material release (Neumann and Rosenheck, 1972) and uptake (Zimmermann et al., 1976). A further important consequence of electroporation is that electroporated membranes are conditioned for fusion if brought into contact (cell electrofusion (Senda et al., 1979; Neumann et al., 1980; Weber et al., 1981; Zimmermann and Scheurich, 1981; Teissié et al., 1982) as well as for insertion of foreign glycoproteins (electroinsertion (Mouneimne et al., 1989)).

Of general interest for molecular biology, gene engineering, therapy, and biotechnology is the direct, electroporative transfer of DNA [electrotransformation, Neumann (1982)] or of other nucleic acids and proteins (Winegar et al., 1989) into recipient cells, microorganisms, and tissues. The main advantage of the electroporative gene transfer is that intact, chemically untreated cell material can be efficiently transfected (Hashimoto et al., 1985; Dower et al., 1988; Wolf et al., 1989). The stable electrotransformation of intact bacteria, yeast, and plant cells is of considerable biotechnological potential (Chassy et al., 1988; Neumann et al., 1989; Chang et al., 1991).

In contrast to the numerous applications, the molecular mechanisms that are operative in the various electroporation phenomena are not well understood. Data analysis and technical optimization strategies are therefore still by and large empirical. Quantitative model calculations only cover partial aspects, such as electric properties (Sowers, 1987; Neumann et al., 1989; Chang et al., 1991). No doubt, further progress in goal-directed applications of the electroporation methods in cell biology, biotechnology, and medicine will greatly benefit from knowledge of the molecular mechanism of membrane electroporation.

For gene delivery, there are general practical guidelines, which are already outlined in the first documentation of electroporative gene transfer leading to cell transformation (Neumann et al., 1982; Förster and Neumann, 1989). In addition, there are detailed technical procedures for many special cases; for example, the Bio-Rad Laboratories Bulletins (1988).

In this discussion of gene transfer by electroporation, emphasis is set on fundamental physical–chemical principles of electrical field effects on cells in general and on membranes and DNA as polyelectrolyte structures in particular. Deeper insight can be gained by inspection and analysis of strength–duration curves, of the electrosensitivity of the cells, as well as of the transformation efficiencies as a function of the concentrations of cells and of DNA. The various concepts and suggestions for goal-directed optimization strategies are outlined primarily with data obtained in our laboratory; the citations are restricted to the original work related to particular aspects of cell electrotransformation. A historical survey on the development and key observations of the electrical field effects on cells is given in Neumann (1989).

HYSTERESIS AND CHEMICAL ELECTROPORATION SCHEME

Membrane electroporation is a new field of biophysical chemistry. There is still a need to classify the observations in terms of physical concepts and to establish an unequivocal physical–chemical terminology. So far, all electroporation data indicate that reversible primary processes and irreversible secondary events can be differentiated and are summarized in Table 8-1, which is a modified version of that in Neumann (1988, 1989).

Electroporation Hysteresis

Transient membrane electroporation can be viewed as a cycle of electrical–chemical membrane events. In the presence of the external field E, there is initiation of pore formation, path $0 \rightarrow A$, and at pulse termination ($E = 0$), pore resealing, path $B \rightarrow 0$ (Fig. 8-1). The state cycle $0 \rightarrow A \rightarrow B \rightarrow 0$ thus represents a relaxation hysteresis (Neumann, 1988; 1989). Crater formation (Neumann, 1989) as observed by electron microscopy (Chang and Reese, 1990) may be classed as an after-field effect also involving osmotic processes.

The structural changes associated with the electroporation hysteresis are most generally described by a degree of poration ξ. The contributions $\tilde{G}(r,E)$ of the free enthalpy, that is, the (reversible) work of electrical interfacial polarization and of the structural changes, can be expressed in terms of a mean pore radius r and the external electrical field strength E. Figure 8-1 shows the cycle of changes in $\tilde{G}(r, E)$ associated with the hysteresis in $\xi(E)$.

The electroporation model of Abidor et al. (1979) and Chernomordik and Chizmadzhev (1989) assumes that during pore formation there is a structural transition from hydrophobic (HO) pores to hydrophilic (HI) pores (Litster, 1975; Weaver and Powell, 1989; Neumann et al., 1991) when a critical pore radius r_c is reached. As seen in Figure 8-1, r_c corresponds

TABLE 8-1. Fundamental Processes of Membrane Electroporation

Physical–Chemical Processes	Electrical Terms
Reversible primary processes	
Primary electric events	
Electric dipole induction and dipole orientation	Dielectrical polarization (Maxwell–Wagner)
Redistribution of mobile ions at phase boundaries (membrane/solution), including	Ionic interfacial polarization (β-dispersion)
a. ionic atmosphere shifts	
b. local activity changes of effectors (e.g., H^+-/pH changes) or Ca^{2+}-ions	
Structural rearrangements	
Conformational changes in protein and lipid molecules	
Phase transitions in lipid domains, resulting in pores, cracks (via pore coalescence), and percolation	Electroporation Electropores, electrocracks, Electropercolation
Annealing and resealing processes (pore closure)	
Energy transfer by conformational cycles	Electroconformational coupling
Nonequilibrium distribution of metabolites	Electropumping
Irreversible secondary processes	
Transient material exchange	Electropermeabilization
Release of cellular compounds (e.g., hemolysis), or of pDNA	Electrorelease Electrocuring
Uptake of external material (e.g., drugs, proteins, antibodies)	Electroincorporation
Transfer of genetic material (e.g., DNA, mRNA, viroids, with stable cell transformation by insertion of foreign genes into the host genome)	Electrotransfection Electrotransformation, Electroporative gene transfer
Cell growth and proliferation	Electrostimulation
Membrane reorganization	
Cell fusion (if membranes are in contact)	Electrofusion
Vesicle formation (budding)	Electrovesiculation Electrobudding
Electromechanical rupture	Dielectric breakdown
Insertion of proteins into membranes	Electroinsertion
Tertiary effects	
Temperature increase due to dissipative processes	Joule heating, dielectric losses
Metal–ion release from metal electrodes	Electroinjection
Electrode surface H and O in _statu nascendi_	Electrolysis

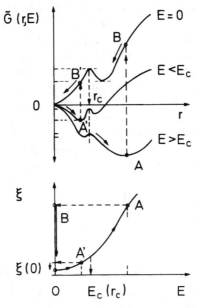

Figure 8-1. The thermodynamics of the electroporation hysteresis. The upper part shows the free enthalpy $\tilde{G}(r, E)$ associated with the electrical and surface work as a function of pore radius r. The lower part shows the corresponding cycle of pore formation and resealing in terms of the degree of poration ξ_p as a function of field strength. In the model of Chernomordik and Chizmadzhev (1989) the structural membrane changes involve a transition from hydrophobic to hydrophilic pores at a critical pore radius r_c; r_c corresponds to a critical value $\xi_{p,c}$, which is reached at a critical value E_c. The majority of the electroporation phenomena appears to require that $E \geq E_c$. Only the electroinsertion of glycoproteins in organelle membranes appears to be on the low-field level of increased hydrophobic pores; that is, at $E \leq E_c$. The sequence of state transitions $0 \to A \to B \to 0$ underlying a supercritical $(E \geq E_c)$ electroporation-resealing cycle is indicated.

to a critical value ξ_c on the lower hysteresis branch. It is, however, not yet known what are the detailed molecular rearrangements in the membrane.

Chemical Electroporation Scheme

The various electroporation phenomena may be described by a general chemical scheme (Neumann and Boldt, 1989, 1990):

$$C \underset{k_R}{\overset{k_p(E)}{\rightleftharpoons}} P \to X \tag{1}$$

In Equation 1, C is the initial membrane state in the absence of external fields. The states P represent a collection of permeabilized (and fusogenic) states of different longevity; $k_p(E)$ and k_R are the (overall) rate coefficients of electroporation and resealing, respectively. If the electroporation hysteresis is coupled to other processes such as material release or uptake, cell lysis, membrane fusion or protein insertion, the states X have to be explicitly specified.

The hysteretic nature of the permeabilization-resealing cycle justifies that the individual rearrangements $C \rightarrow P$, $C \leftarrow P$, and $P \rightarrow X$, respectively, are treated as unidirectional (irreversible) processes. Usually, the reverse transitions $C \leftarrow P$ and some $P \rightarrow X$ processes occur at $E = 0$; they thus represent after-pulse effects. It appears that the transport of DNA through the electroporated membrane can also involve an after-pulse phase of purely diffusive character (Neumann et al., 1982; Neumann, 1989).

ELECTROSENSITIVITY OF CELL POPULATIONS

The electrotransformation of cells, bacteria, and other microorganisms is preferably performed at low, but finite, electrical conductivity. The field effect on the surface membrane is mediated by ionic interfacial polarization and thus requires a finite salt concentration. Joule heating and other tertiary effects (see Table 8-1) are reduced at low ionic strength. Joule heating by itself can, however, be favorable for gene transfer (Wolf et al., 1989). The interpretation of electroporation data of media with low conductivity necessitates an extension of the analytical framework in terms of the solution conductivity (Neumann 1989; Neumann and Boldt, 1989, 1990).

Permeabilization and Recovery

Before the actual electrotransfection experiments are performed, the electrosensitivity of the cells is tested and survival or recovery curves are measured in the absence of added DNA (Wong and Neumann, 1982). Actually, the fraction $f(R)$ of recovered (viable) cells R, $f(R) = R/C_T$, or the percentage $R(\%) = 10^2 \cdot f(R)$ of the total cell number C_T, is plotted as a function of the (initial) field strength at a given pulse duration $t_{P(E)}$, or $R(\%)$ is given as a function of $t_{P(E)}$ at given field strengths. Another variable determining $f(R)$ is the pulse number. The single pulses of a train of pulses are, generally, not equivalent. Because of the longevity (seconds, minutes) of the pulse-induced structural changes in the membrane patches (pole caps) compared with the short pulse durations (0.02–30 msec), a second pulse encounters patches that are not yet resealed. Rotational diffusion will change the position of the cell or of elongated microorganisms, such as a bacterium, relative to the field vector. The analysis of electroporation effects caused by a series of pulses is thus more complicated. In any case, the $f(R)$ correlations will also depend on the composition of the electroporation medium and on the temperature.

If dye staining of the electroporated cells can be measured, as in the case of cells of the green alga *Chlamydomonas reinhardtii*, a particularly efficient procedure can be used to determine the optimum electroporation conditions for transformation or other manipulations (Neumann and Boldt, 1989; Weaver and Powell, 1989). The number of permeabilized cells P is determined in the presence of the staining dye during pulsing (P_0) and at various times after pulse applications. The data in Figure 8-2 not only show that the degree of permeabilization $f(P_0)$, expressed as the percentage of colored cells $G_0(\%) = P_0(\%) = 10^2 \cdot f(P_0)$, is dependent on the medium conductivity, but also that *C. reinhardtii* cells represent a population of individuals of different electrosensitivity.

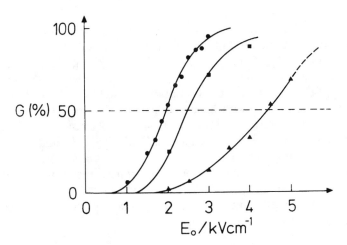

Figure 8-2. Electrosensitivity $G(\%)$ of a cell suspension of the green alga *Chlamydomonas reinhardtii* (wild type 11-32c, Göttingen) to quasirectangular electrical pulses of the initial field intensity E_o and of the pulse length $t_p = 0.2$ msec at different medium conductivities: ● ($\lambda_0 = 3.5 \pm 0.1 \cdot 10^{-4}$ S cm^{-1}), ■ ($\lambda_0 = 1,5 \pm 0,1 \cdot 10^{-4}$ S cm^{-1}), ▲ ($\lambda_0 = 5.6 \pm 0.5 \cdot 10^{-5}$ S cm^{-1}). $G(\%)$ is the percentage of cells that were critically permeabilized such that the (lethal) dye Serva Blue G ($M_r = 854$, largest dimension 2.5 nm) was taken up, thus visibly coloring these cells.

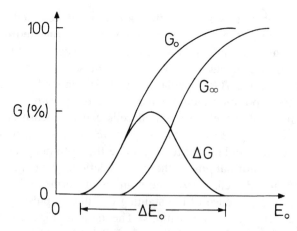

Figure 8-3. Scheme for the optimization of electroporation conditions. $G_0 = P_0(\%)$ is the percentage of cells colored by the dye present during the pulse; $G_\infty = P_\infty(\%)$ is the fraction of cells which did not recover after long resealing times (Neumann and Boldt, 1989, 1990). The difference $\Delta G = G_0 - G_\infty$ represents the transiently permeabilized surviving cells. The optimum field strength range ΔE_0 is shifted when the electroporation conditions are changed.

The after-pulse staining procedure not only permits determination of P_0, but also P_∞, the number of dead permeabilized cells, both as a function of E at a given $t_{P(E)}$ and as a function of $t_{P(E)}$ at given E. Figure 8-3 shows that the difference $\Delta G = \Delta P = P_0 - P_\infty$ has a maximum in the range ΔE_0. There is thus a survival optimum, where further cell manipulations can be performed with highest efficiency. The range ΔE_0 is shifted when the electroporation conditions of ionic strength, temperature, pulse duration, and pulse number are changed. Obviously, ΔE_0 and ΔP must be explored for every particular cell type. In summary, the dye method originally applied as a qualitative tool to estimate the state of electroporated cells (Neumann et al., 1982) can be quantified to yield useful information for an efficient, theoretically guided, optimization of electrical cell manipulations.

Inhomogeneity of Electrosensitivity

The data of Figure 8-2 confirm that, usually, biological cell populations are nonhomogeneous in cell size, state of growth, or their metabolic conditions. Nonspherical cells, such as rod-shaped bacteria in solution, initially have different positions relative to the external electrical field vector.

Because of the predominantly negative surface charges of the cell wall, ionic atmosphere polarization (leading to an induced macrodipole moment) will orient the rods with their longest axis in the direction of the external electrical field. Both oriented bacterial rods and the oriented polyelectrolyte DNA will electrophoretically migrate with different velocity along the electrical field lines. All these factors cause a distribution of critical field strength, E_c, at which a cell starts either to take up a particular dye or a DNA molecule or at which it begins to fuse when brought into contact with other cells. The E_c values for the various cell manipulations, coupled to membrane electroporation, are not necessarily equal. Note that the E_c term discussed in the context of Figure 8-1 refers to a special structural transition.

Therefore, the data in Figure 8-2 strictly represent only the electrosensitivity of C. *reinhardtii* cells with respect to dye uptake in terms of a distribution of critical field strengths E_c. If the field strength E_0 (50%) is taken, where $G(\%) = P_0(\%) = 10^2 \cdot f(P_0) = 50$, as representative for the cell population, a mean value can be defined as

$$\overline{E}_c = E_0 \ (50\%) \tag{2}$$

in terms of a Gaussian distribution (Neumann and Boldt, 1989, 1990). The "width" of the electrosensitivity of a cell population can then be given in terms of a variance $\pm \Delta E_c$. Consistent with the concept of ionic interfacial polarization, preceding the membrane electroporation process, the individual E_c value depends on the size of the individual cell of a given population. For spherical cells, the mean value $\overline{E}_c$ corresponds to a mean value of the cell radius $\overline{a}$, and the variance $\pm \Delta E_c$ reflects a variance $\pm \Delta a$ in the effective cell size.

More complex cell populations may require more than one Gaussian parameter set. In any case, the definition of means such as $\overline{E}_c$ and $\overline{a}$ avoids the ambiguity encountered in the exact practical determination of a threshold field strength as the lowest E_c value of the whole population. It is usually only possible to obtain a range ΔE_{th}, rather than an exact threshold value E_{th} because of the scatter of the experimental data.

Strength – Duration Relationship

A further aspect of enormous practical significance is the field strength–pulse duration relationship, originally discussed in terms of the threshold field strength, E_{th}. Because of the uncertainties in the experimental E_{th} value and because of the distribution reflected in the elec-

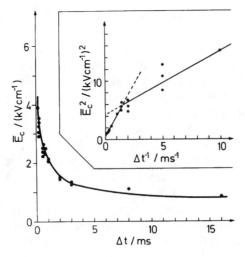

Figure 8-4. Strength–duration relationships of the electroporation of *C. reinhardtii* cells: mean values $\bar{E}_c$ as a function of the pulse duration $\Delta t = t_p$. Insert: Data plot according to the interfacial polarization-flow model $\bar{E}_c^2 = A \cdot t_p^{-1}$, where A is a constant (see Equation 19). Conditions: $\lambda_0 = 1.4\ (\pm 0.4) \cdot 10^{-4}$ S cm^{-1} such that $f(\lambda) = 1$, one rectangular pulse, 293 K; cell density 2×10^7/mL.

trosensitivity curve, the mean value $\bar{E}_c \pm \Delta E_c$ appears to be a more reliable parameter than just the minimum E_c-value of the cell distribution.

Figure 8-4 shows the strength–duration relationship of *C. reinhardtii* cells in terms of the $\bar{E}_c$ values for dye penetration. Since electroporation is initiated by ionic interfacial polarization in conductive media, the energy input $W_{p,c}$ per cell can be expressed as

$$\bar{W}_{p,c} = \text{const} \cdot \bar{E}_c^2 \cdot t_p \tag{3}$$

suggesting a linear relationship between $\bar{E}_c^2$ and t_p^{-1} according to $\bar{E}_c^2 = (\bar{W}_{p,c}/\text{const})t_p^{-1}$.

For media of low conductivity and for large cells, where the relaxation time τ_p of the interfacial polarization is in the same order of magnitude as the pulse duration t_p, an effective $t_{p,\text{eff}}$ must be calculated for the theoretical analysis (see also Equation 19).

The insert in Figure 8-4 shows that for *C. reinhardtii* cells, there are apparently at least two electrosensitivity ranges of the type described by Equation 3. It is not clear whether, in the long pulse range ($t_p = \Delta t \geq 5$

msec), electromechanical cell distortions (Pliquett, 1967) and/or osmotic effects have to be considered or whether $\bar{E}_c$ reaches a saturation value at large t_p, which is independent of t_p.

In any case, the numerical value of a critical field strength (and of a threshold field strength) is a useful information only if the pulse duration of the single pulse application is also given. If pulses can be applied for a long time or if repetitive pulses are applied in series, the threshold for an electroporation phenomenon may be very low, as described by Xie et al. (1990).

MEMBRANE POLARIZATION AND CONDUCTIVITY

As far as the externally applied electrical field E is concerned, membrane electroporation is an indirect field effect. The coupling of E to the membrane structure is mediated by the ionic interfacial polarization. In brief, the electroporation data, including those shown in Figure 8-2, indicate that the field effect is a sequence (Neumann, 1989)

$$E \rightarrow \Delta \varphi \rightarrow \Delta \xi \tag{4}$$

where E induces the transmembrane potential difference, which, in turn, represents a contribution $\bar{E}_m(\Delta \varphi)$ to the mean electrical field forces causing structural rearrangements $(\Delta \xi)$ in the membrane phase (Neumann, 1989).

Ionic Interfacial Polarization

In the case of rectangular pulses, the build up of the induced interfacial transmembrane voltage with time t is described by a linear differential equation; integration yields

$$\Delta \varphi(t) = \Delta \varphi(E) \left[1 - \exp\left(-t/\tau_p \right) \right] \tag{5}$$

where $\Delta \varphi(E)$ is the (time-independent) amplitude and τ_p is the polarization time constant.

Polarization Time Constant

For spherical cells of radius a, the value of τ_p can be estimated from the conductivities of the solution λ_o, membrane λ_m, and cell interior λ_i.

According to Schwan (1957),

$$\tau_p = aC_m(\lambda_i + 2\lambda_o)/[2\lambda_i \cdot \lambda_o + \lambda_m(\lambda_i + 2\lambda_o)] \tag{6}$$

In Equation 6, $C_m \approx 1 \ \mu F \ cm^{-2}$ is the specific membrane capacitance, and $\lambda_m = a \cdot G_m$ is the membrane conductivity (with G_m being the specific membrane conductance). Before the onset of electroporation, the inequality $\lambda_m \ll \lambda_o$ usually applies. In media of low conductivity ($\lambda_o \leq 10^{-4}$ S cm^{-1}), $\lambda_o \ll \lambda_i$ ($\lambda_i \approx 10^{-2}$ S cm^{-1} at 0.1M ionic strength). For this case, Equation 6 takes the simple form

$$\tau_p = aC_m \cdot (2\lambda_o)^{-1}, \tag{7}$$

demonstrating the inverse dependence of τ_p on λ_o. If larger Wien effects are absent, the approximation $\tau = \tau(E = 0)$ and the low-field value $\lambda_o(0)$ may be used. In many electroporation experiments, $\tau_p \ll t_{p(E)}$, such that the constant stationary value $\Delta\varphi(E)$ may be used.

Transmembrane Voltage

For spherical cells of radius $a \gg d$, where d is the membrane thickness, the stationary value of the induced transmembrane potential difference is given by Neumann (1989) as

$$\Delta\varphi(E) = -\tfrac{3}{2}f(\lambda)aE \cdot |\cos \delta|, \tag{8}$$

where δ is the angle between the membrane site considered and the direction of the E vector, and the conductivity factor $f(\lambda)$ is an explicit function of $\lambda_o, \lambda_i, \lambda_m$ and of the ratio d/a (Neumann, 1989). For $\lambda_i, \lambda_o \gg \lambda_m$, $f(\lambda) = 1$. It is important to note that according to Maxwell's definition ($E = -\nabla\varphi$), the electrical potential drops for positive ions in the direction of the external field vector (see Fig. 8-5 and the negative sign in Equation 8).

The membranes of biological cells have natural potential differences ($\Delta\varphi_m$) due to diffusional permselectivities, and $\Delta\varphi_s$ because of an asymmetric distribution of fixed surface charges. The total potential difference, $\Delta\varphi_M$, experienced by a membrane component in the presence of an external field, is given by Neumann (1989) (see Fig. 8-5).

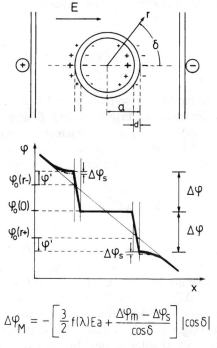

$$\Delta\varphi_M = -\left[\frac{3}{2} f(\lambda) E a + \frac{\Delta\varphi_m - \Delta\varphi_s}{\cos\delta}\right] |\cos\delta|$$

Figure 8-5. Interfacial polarization of a spherical nonconducting shell of thickness d and outer radius a in a constant external field **E**. The stationary electrical potentials are given in polar coordinates of the radius vector **r** and the angle δ, such that the conducting interior of the cell has the constant reference potential $\varphi_0 = 0$. The $\Delta\varphi$ terms are the interfacially induced transmembrane potential differences in the absence of fixed ionic groups and adsorbed ions (surface potential $\Delta\varphi_s = 0$). The dash–point line schematically models the potential profile in the presence of fixed surface charges (here negative).

At the pole caps of spherical cells, that is, at $\delta = 0$ and, hence, $\cos\delta = \pm 1$, the induced transmembrane voltage is a maximum. The maximum stationary value is given by

$$\Delta\varphi_{cap} = -\tfrac{3}{2}f(\lambda)E \cdot a \tag{9}$$

Bacterial Cells

If cell geometries other than spheres have to be considered, ellipsoidal bodies in terms of half-axes can be used. For each half axis a_j, we can write a (modified) Fricke equation for the stationary value in the direction of a_j:

$$\Delta \varphi = -F_j \cdot f_j(\lambda) a_j E \cdot |\cos \delta| \tag{10}$$

where F_j is the form factor. For spheres, $F = 1.5$, a_j is the radius a.

In the case of cylindrical bacteria of length L and of thickness b, the prolate limit $L > b$ may be applied. For bacterial rods oriented with their long axes $(a_L = L/2)$ in the direction of E, the form factor is approximated by $F_L = 1.0$ (Fricke 1953). Therefore, at the pole caps:

$$\Delta \varphi_{cap}(L) = -\tfrac{1}{2} LE \tag{11}$$

for *Escherichia coli* bacteria of an average length of $L = 3.5$ μm and for rectangular pulses of $t_p = 1$ msec, a mean value $\bar{E}_c = 2$ kV cm^{-1} can be estimated from the data of Xie et al. (1990). Inserting this value in Equation 11, yields $\Delta \bar{\varphi} = -0.35$ V (for $t_p = 1$ msec) as the critical transmembrane voltage at which oriented bacteria of mean size start to take up DNA. For longer pulse durations and multiple pulsing $|\Delta \bar{\varphi}|$, and thus $\bar{E}_c$, may have much smaller values. In any case, $|\Delta \varphi| = 0.35$ V is in the order of magnitude 0.2–0.4 V expected for long pulses. It is recalled that for short pulses ($t_p \approx 0.01$–0.1 msec), $|\Delta \varphi| \approx 1$ V (Sale and Hamilton, 1968).

Membrane Conductivity

Data of the type plotted in Figure 8-2 can be used to determine the conductivity λ_m of the membrane, including the cell wall or other envelope structures (Neumann and Boldt, 1989, 1990). If the inequality $\lambda_m \ll \lambda_i, \lambda_o$ holds, the $f(\lambda)$ factor of spherical cells is given by Neumann (1989) as

$$f(\lambda) = \frac{\lambda_o \lambda_i \cdot 2d/a}{(2\lambda_o + \lambda_i)\lambda_m + (2d/a)\lambda_o \lambda_i} \tag{12}$$

For media of low conductivity, $\lambda_o \ll \lambda_i$, Equation 12 reduces to

$$f(\lambda)^{-1} = 1 + [\lambda_m a/(2d)]\lambda_o^{-1} \tag{13}$$

Insertion of Equation 13 into Equation 9 leads to an expression relating $\bar{E}_c$ to λ_o^{-1} by

$$\bar{E}_c = -\frac{2\Delta\bar{\varphi}_c}{3\bar{a}}\left(1 + \frac{\lambda_m \bar{a}}{2d\lambda_o}\right) \tag{14}$$

Equation 14 suggests an evaluation of the minimum value of $(\bar{E}_c)_o = -2\Delta\bar{\varphi}_c/(3\bar{a})$ and thus of $\Delta\bar{\varphi}_c$. The membrane conductivity λ_m is derived from the slope of the graphical data plot as in Figure 8-6. The electroporation data of *C. reinhardtii* cells yield $\bar{E}_c(f = 1) = 1.7$ kV cm^{-1} and $(\lambda_o^{-1})_o = -1,3 \cdot 10^4$ S^{-1} cm. The geometrical "radius" of the most abundant cells is $\bar{a} = 3.5$ μm, and the membrane thickness is taken to be $d = 10^{-6}$ cm.

For the pole cap regions we obtain from Equation 14: $\Delta\bar{\varphi}_{\text{cap},c} = -0.9$ V and $\lambda_m = -(2d)/[\bar{a}(\lambda_o^{-1})_1] = 4.4 \times 10^{-7}$ S cm^{-1}. The critical value of 0.9 V compares well with the estimate of 1 V (Sale and Hamilton, 1968). The dependence of $\bar{E}_c$ on λ_o is quantitatively consistent with the notion of an indirect field effect on the membrane structure; pore formation is mediated by interfacial polarization. At low solution conductivity, λ_m cannot be neglected; instead of $f(\lambda) = 1$, Equation 13 has to be applied.

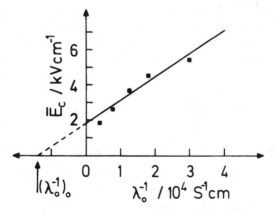

Figure 8-6. The dependence of the mean critical field strength, $\bar{E}_c$, for dye uptake by electroporation of *C. reinhardtii* cells on the solution conductivity λ_0 at 293°K, according to Equation 14 of the text. See Fig. 8-2 caption. Data evaluation yields the mean critical transmembrane voltage $\Delta\varphi_c = -0.9$ V and the membrane conductivity $\lambda_m = 4.4 \cdot 10^{-7}$ S cm^{-1}.

PULSE SHAPE AND DATA COMPARISON

Frequently, electroporation data that were obtained with different pulse shapes are compared. An adequate comparison requires a deeper insight into the details of the membrane polarization processes.

The ion accumulations at the membrane interfaces, primarily at the pole caps (Fig. 8-5) can be described in terms of an interfacial polarization current, I_p, and a corresponding surface polarization conductance, G_p. The electric part W_p of the free enthalpy change or (reversible work) for the induction of the transmembrane voltage, $\Delta\varphi$, and subsequent pore formation is then expressed as

$$W_p = \int_0^t I_p |\Delta\varphi|\, dt \tag{15}$$

In the ohmic region, where G_p in $I_p = G_p \cdot (-\Delta\varphi)$ is a constant, we obtain

$$W_p = G_p \int_0^t (\Delta\varphi)^2\, dt \tag{16}$$

Rectangular Pulses

If the pulse duration t_p and the polarization time constant τ_p are comparable, W_p is a function of both τ_p and t_p. Substitution of Equation 5 into Equation 16 and integration in the boundaries $t = 0$ and $t = t_p$ yields

$$W_p = G_p [\Delta\varphi(E)]^2 \cdot t_{p,\text{eff}} \tag{17}$$

where the effective pulse duration is given by

$$t_{p,\text{eff}} = t_p - \frac{\tau_p}{2}(3 + e^{-2t_p/\tau_p} - 4e^{-t_p/\tau_p}) \tag{18}$$

Often, cell size and ionic content are such that $\tau_p \ll t_p$. For this limit, $t_{p,\text{eff}} = t_p$ and $W_p = G_p \cdot [\Delta\varphi(E)]^2 \cdot t_p$.

At the cap regions of spherical cells, where Equation 9 holds, the mean polarization energy $\overline{W}_{p,c}$ of a cell distribution ($\overline{a} \pm \Delta a$) is given by

$$\overline{W}_{p,c} = G_p \left[\frac{3\overline{a}}{2} f(\lambda) \right]^2 \cdot \overline{E}_c^2 \cdot t_p \tag{19}$$

thus specifying Equation 3 in more detail.

Condenser Discharge Pulses

The exact analysis of exponential pulse intensities is more elaborate. Condenser discharge pulses of the form

$$E(t) = E_o \cdot e^{-t/\tau_E} \tag{20}$$

where $\tau_E = R'C'$, R' and C' are the resistance and the capacitance, respectively, of the discharge circuit [see Neumann (1989) for more details], induce a transmembrane voltage of

$$\Delta \varphi(t) = \Delta \varphi(E_0) \frac{\tau_E}{\tau_E - \tau_p} (e^{-t/\tau_E} - e^{-t/\tau_p}) \tag{21}$$

it is assumed that τ_p is field independent [see Neumann (1989)]. Substitution of Equation 21 into Equation 16 yields

$$W_p = G_p \cdot \left[\Delta \varphi(E_0)^2 \right] \cdot \left(\frac{\tau_E}{\tau_E - \tau_p} \right)^2 \cdot X \tag{22}$$

where

$$X = \int_0^t (e^{-t/\tau_E} - e^{-t/\tau_p})^2 \, dt$$

$$= -\frac{\tau_E}{2} (e^{-2t/\tau_E} - 1) - \frac{\tau_p}{2} (e^{-2t/\tau_p} - 1)$$

$$+ \frac{2\tau_E \tau_p}{\tau_E - \tau_p} (e^{-t(\tau_E + \tau_p)/\tau_E \cdot \tau_p} - 1) \tag{23}$$

Quasirectangular Pulses

For quasirectangular (QR) pulses, the upper integration boundary is $t = t_p$. In the case of very rapid membrane polarization, that is, $\tau_p \ll \tau_E$, we obtain from Equations 21 to 23, respectively,

$$\Delta\varphi(t) = \Delta\varphi(E_0)e^{-t/\tau_E} \tag{24}$$

and

$$W_p = G_p[\Delta\varphi(E_0)]^2 \cdot \frac{\tau_E}{2}(1 - e^{-2t/\tau_E}) \tag{25}$$

For QR pulses, it is taken that $t = t_p$.

Purely Exponential Pulses

In the case of purely exponential (CD) pulses, the limit $t \to \infty$ applies and the polarization energy is given by

$$W_p = G_p[\Delta\varphi(E_0)]^2 \cdot \frac{\tau_E}{2} \tag{26}$$

Data Comparison

An adequate comparison of electroporation data can now be based on the condition of equal polarization energy. For instance, a comparison can be made of data obtained from application of rectangular pulses and CD pulses for the conditions $\tau_p \ll t_p$ and $\tau_p \ll \tau_E$, respectively. Hence Equations 19 and 26 apply and the mean values of the cell population can be obtained:

$$\bar{E}_c^2 \cdot t_p = (\bar{E}_0)_c^2 \tau_E/2 \tag{27}$$

If the condition of equal initial field strength is selected, that is, $\bar{E}_c = (\bar{E}_0)_c$, a reasonable comparison requires that $\tau_E = 2 \cdot t_p$. Alternatively, selecting the condition $t_p = \tau_E$, the equality of $(\bar{E}_0)_c = \bar{E}_c \cdot 2^{1/2}$ is required for the comparison.

AC Pulses

For the case of AC pulses of angular frequency $\omega = 2\pi\nu$ (ν being the frequency) and of duration t_p, the term $[\Delta\varphi(E)]^2$ in Equation 17 has to be replaced by the effective peak value $([\Delta\hat{\varphi}(\hat{E})]^2/2)/(1 + w^2\tau_p^2)$. For low frequencies $\omega \ll \tau_p^{-1}$, by substitution in Equation 19, values are obtained for a spherical cell distribution

$$\overline{W}_{p,c} = G_p\left(\frac{3\bar{a}}{2}f(\lambda)\right)^2 \cdot \frac{\left(\hat{\bar{E}}_c\right)^2}{2} \cdot t_p \tag{28}$$

The comparison, for instance with data from CD pulses, must be based on the equality

$$\left(\hat{\bar{E}}_c\right)^2 \cdot t_p = \bar{E}_{o,c}^2 \cdot \tau_E \tag{29}$$

In any case, W_p will reach a limiting value $W_{p,c}$, independent of t_p, when the electroporated membrane starts to conduct ions through the pores appreciably.

PHYSICAL CHEMISTRY OF DNA ELECTRODELIVERY

Electroporative DNA transfer is a multiphase process, favored by adsorption of DNA on the cell surface (Neumann and Boldt, 1990; Neumann et al., 1991; Xie et al., 1990). Beside direct adsorption to the cell state C [see Scheme 1], there are certainly transient contacts of DNA with cell surface structures such as cell walls and with the plasma membrane. The transport of DNA through the cell wall and through the electroporated membrane is also determined by the polyelectrolyte nature of DNA and by the cell surface charges.

Cell Surface Contact of DNA

To describe the interaction of DNA with cell surfaces in DNA electrodelivery, the state hysteresis $C \rightleftarrows P$ (Scheme 1) can be coupled to association–dissociation processes of the DNA (D) and to the (electro-)diffusion of DNA through the electroporated patches (P) into the cell interior (Neumann et al. 1991). In Scheme 30,

$$D + mC \rightleftharpoons D \cdot C \qquad (30)$$
$$\updownarrow \qquad\qquad \updownarrow$$
$$D + mP \rightleftharpoons D \cdot P \leftrightarrow P \cdot D \rightarrow D_{in} \rightarrow TC$$

The states $D \cdot C$ and $D \cdot P$ denote the surface-bound DNA, $D \cdot P \leftrightarrow P \cdot D$ represents the transmembrane diffusion of DNA, D_{in} is the DNA that entered the cytoplasma (transfection), TC is the final state of the transformed cell, and m is the maximum number of surface-(binding) sites for DNA per cell.

The data analysis can now be based on the assumption that the transformation probability increases with the increase in the concentration of the transient contact $D \cdot P$. In the regime where we may assume that the concentration of transfected cells $[TC]$ is given by the proportionality $[TC] \sim [D \cdot P]$, the degree of transfection ξ_T is given by

$$\xi_T = \frac{T}{T_{max}} = \frac{[TC]}{[TC]_{max}} = \frac{[D \cdot P]}{[D \cdot P]_{max}} \qquad (31)$$

where T is the number of transfected cells and T_{max} the maximum value.

If mass action laws are applied to the stationary states of DNA binding to m independent sites, we obtain for Scheme 30:

$$\frac{[D \cdot P]}{[D \cdot P]_{max}} = \frac{[D]}{[D] + \bar{Q}(D)} = \frac{m[C]}{m[C] + \bar{Q}(C)} \qquad (32)$$

In Equation 32, the distribution constants $\bar{Q}(D)$, in terms of DNA concentration, and $\bar{Q}(C)$, in terms of cell surface sites, should be the same. Thus $\bar{Q}(D) = \bar{Q}(C) = \bar{Q}$. Hence,

$$\bar{Q} = [D]\frac{m[C] + m[P]}{[D \cdot C] + [D \cdot P]}$$

$$= K_1\frac{(1 + K_0)}{(1 + K_0')} \qquad (33)$$

where the individual distribution constants are defined by $K_0 = [P]/[C]$, $K_1 = [D] \cdot m[C]/[D \cdot C]$, $K'_0 = [D \cdot P]/[D \cdot C]$, and $K_2 = K_1 K_0 / K'_0$.

Now the Equations 31 and 32 are combined and reorganized to yield the practically useful expression

$$\frac{1}{T} = \frac{1}{T_{max}}\left(1 + \bar{Q} \cdot \frac{1}{[D]}\right) \tag{34}$$

For the graphical data evaluation, note that $[D]$ in Equation 34 refers to free, unbound DNA. If, however, T^{-1} is plotted versus $[D_T]^{-1}$, the total concentration $[D_T]$ of the DNA applied, the parameters T_{max} and $\bar{Q}$ can be readily estimated from the range of high $[D_T]$ values to low $[D_T]^{-1}$-values. For instance, the transformation of *Corynebacterium glutamicum* is dependent on the concentrations of DNA and cells (Wolf et al., 1989). Data evaluation according to Equation 34 is shown in Figure 8-7 (Neumann et al., 1991). We obtain $\bar{Q} = 2(\pm 1) \cdot 10^{-9}$ M (DNA), $m = 25$,

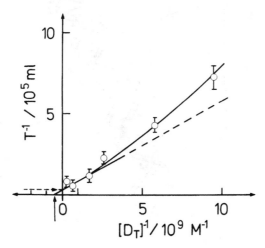

Figure 8-7. Electrotransformation of intact *C. glutamicum* cells at a density of 10^{10}/mL at 4°C. Quasirectangular pulse, $E_0 = 10$ kV cm^{-1}, $t_p = 5$ msec; medium: 0.272 M sucrose, 0.01 M Hepes buffer, 1 mM $MgCl_2$, KOH adjusted to pH 7.4 at 4°C; pUL-330-plasmid DNA (5.2 kb). The Langmuir adsorption-type relationship between the number of transformants, T per milliliter, and the total concentration of DNA $[D_T]$, is represented in terms of Equation 34. The dashed line refers to the linear dependence on $[D]^{-1}$; M(DNA) $= 3.43 \cdot 10^6$ g/mol.

$T_{max} = 3.3 \times 10^5$ mL^{-1} at a cell density of 10^{10} cell/mL, for quasirectangular pulses ($E_0 = 10.5$ kV cm^{-1}, $t_p = 5$ msec).

Orientation of DNA and Bacterium Rods

Since DNA is a linear polyelectrolyte, an applied electrical field E displaces the counterion atmosphere relative to the polyanionic DNA rod, orienting the induced-dipolar rod with its long axis in the direction of E (Eigen and Schwarz, 1962). Figure 8-8 shows how counterion flow along the DNA induces a stationary flow dipole that orientates.

A superhelical plasmid DNA can also be considered as an overall rod, which, in an applied electrical field, is orientated. The rotation time

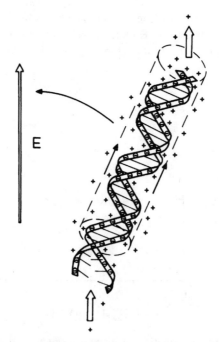

Figure 8-8. Ionic atmosphere flow polarization of DNA induced by counterion displacement along the helix axis in an external electric field **E**. The stationary-induced dipole moment is due to counterion deficiency at the unscreened phosphate groups of the lower end and to counterion accumulation at the upper end of double helix (Werner, 1989).

constant is dependent on E and on the length of the rotation axis (Eigen and Schwarz, 1962).

For the plasmid PBRN 3 (2 kb), the orientation time constant is about 5 μsec at $E = 20$ kV cm^{-1}, 293 K (Werner, 1989). If the pulse duration is in the millisecond time range, the DNA orientation is very rapid. At low ionic content of the solution, however, plasmid DNA may uncoil (Werner, 1989).

Cell surface structures, such as a wall and the outer side of the plasma membrane, are usually net negatively charged and are therefore surrounded by a net positively charged counterion atmosphere. Rod-like cells, such as bacteria, are similarly flow polarized like DNA and orient with their long axis in the direction of the external field (Fig. 8-9), provided the pulse duration t_p is large enough compared to the rotational time constant (estimated to be in the order of milliseconds for bacteria such as *E. coli* with $L \approx 3$ μm in fields of the order of several kilovolts per centimeter).

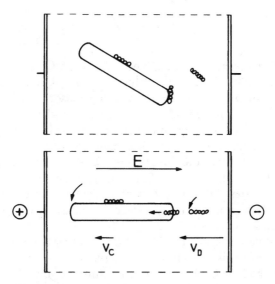

Figure 8-9. Model for the interaction of superhelical plasmid DNA and bacterium rods in an applied electric field **E**. Upper part: $E = 0$; some DNA is adsorbed to the bacterial surface. At large fields, there can be complete orientation of the long axes of the particles in the direction of **E** and electrophoretic movement of the anionic particles. Since $v(D) > v(C)$, DNA can presumably be electrophoretically drawn through one of the electroporated pole caps of the bacterium.

Electrodiffusion of DNA and of Cells

As a whole, cell surfaces are, like DNA, polyanionic. Elongated cells, such as bacteria, and DNA not only orientate in the presence of an external field, but, as polyanions, they move electrophoretically in a drift direction opposite to the direction of the external field vector (Fig. 8-9).

Generally, the velocity vector $\mathbf{v}_i$ of a particle i is related with the field vector $\mathbf{E}$ through

$$\mathbf{v}_i = (z_i/|z_i|)u_i\mathbf{E} \tag{35}$$

where z_i is the charge number (with sign) and u_i is the electrical mobility of the particle i. For anionic particles, $(z_i/|z_i|) = -1$. Therefore, the electrophoretic movement of DNA (D) and of bacterial cells (C) is described, respectively, by

$$\mathbf{v}(D) = -u(D)\cdot\mathbf{E}$$

$$\mathbf{v}(C) = -u(C)\cdot\mathbf{E} \tag{36}$$

It is known that the mobility of DNA, $u(D) \approx 1.5\ (\pm 0.5)\ 10^{-4}\ \mathrm{cm}^2\ \mathrm{V}^{-1} \mathrm{sec}^{-1}$ (Werner, 1989; Eigen and Schwarz, 1962) is independent on the chain length. Generally, the electric mobility tensor u_i of (geometrically anisotropic) particles is dependent on the surface charge density and on particle size (and on the direction of the longest particle axis relative to $\mathbf{E}$). Cell surface structures are usually less charged than DNA and cells are larger than DNA; therefore, $u(D) > u(C)$. As a consequence, in the presence of a field $\mathbf{E}$, DNA moves faster than larger cells (Fig. 8-9). Hence, it is conceivable that DNA is electrophoretically inserted in, and drawn through, the electroporated membrane cap facing the cathodic electrode of the measuring chamber.

Electroporative bacterial transformation requires large pulse durations. The reason is, most likely, the network structure of the porous cell wall. Externally applied DNA has to migrate through the wall components (equivalent to a higher solution viscosity) (electro-) diffusively before entering the plasma membrane.

Finally, it should be realized that the transformed cells T or the degree of transformation $\xi_T = T/T_{\max}$ reflects the last member of the sequence

$$E \rightarrow \Delta\varphi(E) \rightarrow \Delta\xi_p(E) \rightarrow \Delta\xi_T \tag{37}$$

Therefore, ξ_T is likely to depend on E and t_p via the degree of poration ξ_p in a complicated manner. Nevertheless, parameters related to ξ_T and ξ_p, as discussed here, are useful for studies aimed at elucidating the mechanisms of membrane electroporation and of gene electrodelivery.

SUMMARY

Direct gene delivery mediated by electric pulses is described in physical chemical terms by an electroporation-resealing hysteresis coupled to electrodiffusive migration of DNA through cell wall structures and electroporated plasma membranes. Deeper insight into the various electroporation phenomena like electrotransfection, electrofusion and electroinsertion, is gained from inspection of the electrosensitivity or recovery curves of cell populations as well as by analysis of pulse intensity-duration relationships. The results indicate possible mechanisms and are instrumental for goal-directed optimization strategies of practical applications of electroporation techniques.

ACKNOWLEDGMENTS

I thank my co-workers Drs. E. Boldt, H. Wolf, and E. Werner and Mrs. M. Pohlmann for careful typing of the manuscript. Financial support of the Deutsche Forschungsgemeinschaft Grant Ne 227/g-2 and the Fonds der Chemie is gratefully acknowledged.

References

Abidor, I. G., Arakelyan, V. B., Chernomordik, L. V., Chizmadzhev, Y. A., Pastushenko, V. F., and Tarasevich, M. R. (1979). Electric breakdown of bilayer lipid membranes. I. The main experimental facts and their qualitative discussion. Bioelectrochem. Bioenerg. **6**: 37–52.

Bio-Rad Laboratories, Bulletins (1988). Richmond, CA.

Chang, D. C., Chassy, B. M., Saunders, J. A., and Sowers, A. E., eds. (1991). Handbook of Electroporation and Electrofusion. Academic Press, Orlando.

Chang, D. C., and Reese, T. S. (1990). Changes in membrane structure induced by electroporation as revealed by rapid-freezing electron microscopy. Biophys. J. **58**: 1–12.

Chassy, B. M., Mercenier, A., and Flickinger, J. (1988). Transformation of bacteria by electroporation. TIBTECH. **6**: 303–309.

Chernomordik, L. V., and Chizmadzhev, Y. A. (1989). Electrical breakdown of lipid bilayer membranes: Phenomenology and mechanism. Pages 83–95. In Electroporation and Electrofusion in Cell Biology. Neumann, E., Sowers, A. E., and Jordan, C., eds. Plenum Press, New York.

Dower, W. J., Miller, J. F., and Ragsdale, C. W. (1988). High efficiency transformation of E. coli by high voltage electroporation. Nucleic Acids Res. **16**(13): 6127–6145.

Eigen, M., and Schwarz, G. (1962). Structure and kinetic properties of polyelectrolytes in solution, determined from relaxation phenomena in electric fields. Pages 309–335. In Electrolytes. Pesce, B., ed., Pergamon Press, Oxford.

Fricke, H. (1953). The electric permittivity of a dilute suspension of membrane-covered ellipsoids. J. Appl. Phys. **24**: 644–646.

Förster, W., and Neumann, E. (1989). Gene transfer by electroporation: A practical guide. Pages 299–318. In Electroporation and Electrofusion in Cell Biology. Neumann, E., Sowers, A. E., and Jordan, C., eds., Plenum Press, New York.

Hashimoto, H., Morikawa, H., Yamada, Y., and Kimura, A. (1985). A novel method for transformation of intact yeast cells by electroinjection of plasmid DNA. Appl. Microbiol. Biotechnol. **21**: 336–339.

Litster, J. D. (1975). Stability of lipid bilayers and red blood cell membranes. Phys. Lett. **53A**: 193–194.

Mouneimne, Y., Tosi, P. F., Gazitt, Y., and Nicolau, C. (1989). Electroinsertion of xeno-glycophorin into the red blood cell membrane. Biochem. Biophys. Res. Commun. **159**: 34–40.

Neumann, E. (1988). The electroporation hysteresis. Ferroelectrics **86**: 325–333.

Neumann, E. (1989). The relaxation hysteresis of membrane electroporation. Pages 61–82. In Electroporation and Electrofusion in Cell Biology. Neumann, E., Sowers, A. E., and Jordan, C., eds. Plenum Press, New York.

Neumann, E., and Boldt, E. (1989). Membrane Electroporation: Biophysical and biotechnical aspects. Pages 373–382. In Charge and Field Effects in Biosystems-2. Allen, M. J., Cleary, S. F., and Hawkridge, F.M., eds. Plenum Press, New York.

Neumann, E., and Boldt, E. (1990). Membrane electroporation: The dye method to determine the cell membrane conductivity. Pages 69–83. In Horizons in Membrane Biotechnology, Progress in Clinical and Biological Research, Vol. 343. Nicolau, C., and Chapman, D., eds. Wiley-Liss, New York.

Neumann, E., Gerisch, G., and Opatz, K. (1980). Cell fusion induced by high electric impulses applied to Dictyostelium. Naturwissenschaften **67**: 414–415.

Neumann, E., and Rosenheck, K. (1972). Permeability changes induced by electric impulses in vesicular membranes. J. Membr. Biol. **10**: 279–290.

Neumann, E., and Rosenheck, K. (1973). Potential difference across vesicular membranes. J. Membr. Biol. **14**: 194–196.

Neumann, E., Schaefer-Ridder, M., Wang, Y., and Hofschneider, P. H. (1982). Gene transfer into mouse lyoma cells by electroporation in high electric fields. EMBO J. **1**: 841–845.

Neumann, E., Sowers, A. E., and Jordan, C. A., eds. (1989). Electroporation and Electrofusion in Cell Biology. Plenum Press, New York.

Neumann, E., Sprafke, A., Boldt, E., and Wolf, H. (1991). Biophysical Digression on Membrane Electroporation. Pages 77–90. In Handbook of Electroporation and Electrofusion. Chang, D. C., Chassy, B. M., Saunders, J. A., and Sowers, A. E., eds. Academic Press, Orlando.

Pliquett, F. (1967). Das Verhalten von Oxitrichiden unter dem Einfluß des elektrischen Feldes. Z. Biol. **116**: 9–22.

Sale, A. J. H., and Hamilton, W. A. (1968). Effects of high electric fields on microorganisms. III. Lysis of erythrocytes and protoplasts. Biochim. Biophys. Acta **163**: 37–43.

Schwan, H. P. (1957). Electrical properties of tissue and cell suspensions. Adv. Biol. Med. Phys. **5**: 147–209.

Senda, M., Takeda, J., Shunnosuke, A., and Nakamura, T. (1979). Induction of cell fusion of plant protoplasts by electrical stimulation. Plant Cell Physiol. **20**: 1441–1443.

Sowers, A. E., ed. (1987). Cell Fusion, Plenum Press, New York.

Teissié, J., Knutson, V. P., Tsong, T. Y., and Lane, M. D. (1982). Electric pulse-induced fusion of 3T3 cells in monolayer culture. Science **216**: 537–538.

Weaver, J. C., and Powell, K. T. (1989). Theory of Electroporation. Pages 111–126. In Electroporation and Electrofusion in Cell Biology. Neumann, E., Sowers, A. E., and Jordan, C., eds. Plenum Press, New York.

Weber, H., Förster, W., Jacob, H.-E., and Berg, H. (1981). Microbiological implications of electric field effects. Z. Allg. Mikrobiol. **21**: 555–562.

Werner, E. (1989). Ion pairing and flexibility of polyelectrolytes, linear, and superhelical pDNA. PhD Thesis, University of Bielefeld, Germany.

Winegar, R. A., Phillips, J. W., Youngblom, J. H., and Morgan, W. F. (1989). Cell electroporation is a highly efficient method for introducing restriction endonucleases into cells. Mutation Res. **225**: 49–53.

Wolf, H., Pühler, A., and Neumann, E. (1989). Electrotransformation of intact and osmotically sensitive cells of Corynebacterium glutamicum. Appl. Microbiol. Biotechnol. **30**: 283–289.

Wong, T. K., and Neumann, E. (1982). Electric field mediated gene transfer. Biochem. Biophys. Res. Comm. **107**: 584–587.

Xie, T.-D., Sun, L., and Tsong, T. Y. (1990). Study of mechanisms of electric field-induced DNA transfection: DNA entry by surface binding and diffusion through membrane pores. Biophys. J. **58**: 13–19.

Zimmermann, U., Pilwat, G., and Riemann, F. (1974). Reversibler dielektrischer Durchbruch von Zellmembranen in elektrostatischen Feldern. Z. Naturforsch. **29c**: 394–395.

Zimmermann, U., Riemann, F., and Pilwat, G. (1976). Enzyme loading of electrically homogeneous human red blood cell ghosts prepared by dielectric breakdown. Biochim. Biophys. Acta **436**: 460–474.

Zimmermann, U., and Scheurich, P. (1981). High frequency fusion of plant protoplasts by electric fields. Planta **151**: 26–32.

Zimmermann, U., Schulz, J., and Pilwat, G. (1973). Transcellular ion flow in *Escherichia coli B* and electrical sizing of bacteria. Biophys. J. **13**: 1005–1013.

CHAPTER 9

Clinical Applications of Electroporation

S. B. Dev

G. A. Hofmann

ABSTRACT

Electroporation is a well-established physical technique of introducing molecules, such as drugs, antibodies, or DNA, into cells by creating transient pores in the cell membrane. This is done by applying an electrical field to a suspension of cells with the transfectant (e.g., drugs) in a container with electrodes. Applications include preparation of gene libraries and genetic manipulation. Recently, however, the technology has gone beyond research laboratories and is being applied in clinical sciences. In this chapter applications are discussed both in the areas of drug delivery and of gene therapy. Examples include cancer, AIDS, restenosis, hemophilia B, and aplastic anemia, the last two nearing clinical trial. The advantages of electroporation over retrovirus-based gene therapy are discussed, and it is speculated that all single gene diseases with a genomic size less than 190 kb should be amenable to electroporetic gene therapy. Electroinsertion, the related technique of creating small pores to insert molecules in the surface of the cell membrane, may provide an effective way to overcome the current problems of delivering peptides and proteins.

INTRODUCTION

The technology of electroporation, particularly for drug delivery and gene therapy, will find many more applications in research, leading to clinical use in the future, e.g., in gene silencing, gene activation, production of recombinant proteins for therapeutic purpose, and multiple drug resistance gene which plays a major role in cancer chemotherapy.

CURRENT APPROACHES TO DRUG DELIVERY

The literature on drug delivery is extensive (Langer, 1990; Gregoriadis and Florence, 1993; Wallace and Laskar, 1993), covering (1) mechanism of drug release, (2) advantages of controlled release, and (3) problems associated with drug delivery that include limited target access, toxicity for healthy cells, and loss of efficacy in transit. New approaches to delivery, such as liposome encapsulation, chemical modification of the parent drug, covalent linkage to antibody, pumps and polymers, and electric-field mediated delivery [iontophoresis (IP) and electroporation (EP)], are also discussed.

In this review, the main emphasis will be on EP, although the basic differences between IP and EP are summarized. Both ex vivo and in vivo applications are discussed. Delivery of peptide and protein drugs pose special problems because of their very low bioavailability by conventional methods of delivery (Wallace and Lasker, 1993). The technique of electroinsertion may overcome this difficulty.

Electrical Field-Mediated Drug Delivery, IP versus EP

Iontophoretic drug delivery has been in use for a long time. Most of the studies have been on transdermal delivery, with or without absorption enhancers. Both continuous and pulsed DC fields have been used. Despite many publications on iontophoretic delivery of drugs, the progress has been slow with macromolecules. Differences between IP and EP are substantial. These are as follows:

1. EP creates many new routes of administration for the species to be delivered. The mechanism for this is not well understood, but it is believed that this occurs because of change in conformation in the lipid bilayer. In contrast, IP uses only available routes for the drug. For transdermal delivery this would mean hair follicles and sweat glands.

2. Generally, in the IP delivery mode, the possibility exists of tissue damage above an iontophoretic potential of 10 V, the corresponding

amplitude in the EP mode can be as large as 1000 V without any such damage.

3. EP allows incorporation of molecules into the cell cytoplasm. Abraham et al. (1993) have reported that the flux of a therapeutic molecule, such as LHRH, through the human skin can be increased by an order of magnitude if a single electroporation pulse is followed by DC, compared to DC alone or the absence of an electrical field. Thus, EP shows great potential in delivering, non-invasively, an important class of hormones.

Electroporetic Drug Delivery: Clinical Applications

Recently, we described how rapidly EP has progressed from research laboratories to applications in the clinical arenas (Hofmann and Dev, 1993). Studies on cancer and AIDS have completed Phase I clinical trials. In vivo animal models have been developed for prevention of restenosis, reclosure of arteries, that occur in 35–50% of the patients who undergo balloon angioplasty.

Electrochemotherapy: A Novel Cancer Treatment

Electrochemotherapy (ECT) combines electroporation with chemotherapeutic agents to treat cancers. This mode of treatment was first introduced in animal models by Okino and Mohri (1993), who have since then experimented with a large number of rats, treating both hepatocellular and Lewis lung carcinomas (Okino et al., 1992 and the references therein). The exact order of ECT is important—systemic administration of anticancer drug followed by delivery of electrical field pulses at the site of the cancer.

The rationale behind this approach is the following. Some cytostatic drugs like bleomycin (BLM) are very poorly permeable into the tumor cells. To overcome this problem, the drug can be injected intramuscularly, intravenously, or subcutaneously and the tumor nodules subjected to a single or a series of short and intense electrical field pulses. This treatment will result in considerable increase in the uptake of the drug.

The maximum killing effect in the tumors was achieved when both drug (D) and an electrical field (E) were used; that is, $D + E +$, compared to the three controls of $D - E -$, $D - E +$, and $D + E -$. After a single treatment, it was found that there was a 47% reduction in the tumor size at the end of four days, whereas the tumor continued to grow in the control groups. Kanesada (1990), has also treated metastatic tumors using other anticancer drugs, such as peplomycin, cisplatin, mitomycin C, and cyclophosphamide and has shown reduction in the number of metastatic

TABLE 9-1. Number of Lung Metastasis on the 14th Day After Treatment

Group	No. of Mice	Median (Range) No. of Lung Metastasis
Control	8	36 (20–56)
HVP[a]	8	29 (10–55)
PEP[b]	8	27 (18–41)
PEP + HVP	8	4 (0–10)

[a] HVP, high voltage pulse.
[b] PEP, peplomycin.

tumors with $D + E +$. Tables 9-1 and 9-2 summarize the results of Kanesada (1990).

Soon after the findings of Okino et al. (1992), Mir and his team initiated systematic experiments on electropermeabilization, starting with in vitro work on different cell lines. This was followed by in vivo ECT in animals and, subsequently, the first clinical trial with seven patients with head and neck squamous cell carcinoma involving 34 tumor nodules (Mir et al., 1988, 1991a, 1991b; Belehradek et al., 1991). An example of what caliper electrodes would look like in practice is shown in Figure 9-1. Mir et al. (1992) also extended their studies to combine ECT with subsequent injection of very low-dose IL-2 over a few days, which achieved a much higher rate of completely cured animals.

TABLE 9-2. Effect of ECT on the Reduction of Number of Lung Metastasis with Various Anticancer Drugs ($p < 0.01$)

Group	No. of Mice	Median (Range) No. of Lung Metastasis
Control	5	56 (30–83)
CPA[a]	5	0 (0–1)
MMC	5	21 (9–34)
CDDP	5	55 (41–72)
CPA + HVP[b]	5	0 (0)
MMC[c] + HVP	5	6 (2–11)
CDDP[d] + HVP	5	5 (0–11)

[a] CPA, Cyclophosphamide.
[b] HVP, high voltage pulse.
[c] MMC, Mitomycin C.
[d] CDDP, Cisplatin.

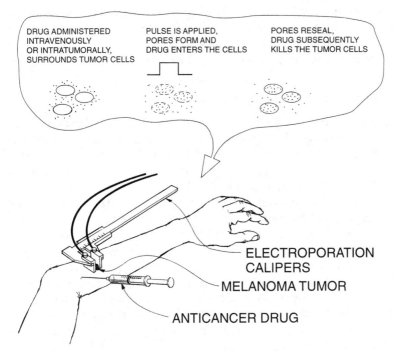

| DRUG ADMINISTERED INTRAVENOUSLY OR INTRATUMORALLY, SURROUNDS TUMOR CELLS | PULSE IS APPLIED, PORES FORM AND DRUG ENTERS THE CELLS | PORES RESEAL, DRUG SUBSEQUENTLY KILLS THE TUMOR CELLS |

ELECTROPORATION CALIPERS
MELANOMA TUMOR
ANTICANCER DRUG

Figure 9-1. Schematic diagram of the ECT process and treatment of melanoma.

In in vitro electroporation experiments with cell suspensions and different anticancer drugs, the cytotoxic effect was found to be maximum with bleomycin compared with other drugs. In addition, the effect was several hundred-fold to several thousand-fold greater (depending on the cell lines) for electroporated cells than with non-electroporated cells. With animal experiments, Mir et al. (1991b) showed that a dose of only 0.5 μg of bleomycin, followed by electroporation, was able to retard tumor growth in animals. This is the equivalent of 1 mg/day for five days without electroporation, a reduction of 10,000-fold. No necrosis has been found following ECT experiments.

In clinical experiments involving patients with head and neck squamous cell carcinoma, Mir et al. (1991a) used stainless-steel strips, pressed against the nodules on the surface of the skin, to apply 8×100 μs pulses from a square-wave generator. The results of this trial were as follows: partial regression 9; complete regression 14; growth retarded with respect to the very rapid increase in the untreated nodules: 6; no change observed: 2; and results not given: 3. Side effects following ECT treatment were minimal; some muscle contraction at the site of the treatment (but no pain) that disappeared immediately after the treatment terminated. Slight oedema that was present also disappeared after 12 and 48 hr.

Electroinsertion and its Application to AIDS

Electroinsertion (EI) is a technique that allows insertion of proteins into the cell membrane by application of an electrical field that is just below the threshold necessary to create normal pores (Mouneimne et al., 1992). Previous to the EI technology, attempts have been made to cure AIDS by soluble CD4 (sCD4). CD4, a recaptor for HIV, is a glycoprotein found on the surface of T4 lymphocytes. After binding to CD4, HIV enters the T4 cells and replicates. The attack on the T4 cells severely depletes their numbers, whereas the number of HIV particles in the blood stream increases and ultimately overwhelms the immune system, leading to severe immunodeficiency and opportunistic infections. The approach of therapeutic CD4 administration has not been successful for three reasons. Specifically, (1) the half-life of sCD4 is only 10 min in the blood circulation. (2) the binding affinity of sCD4 to HIV on T4 cells is high and it tends to disrupt the viral envelope, and (3) the amount of sCD4 needed for a therapeutic dose is too high.

To deal with these problems, Nicolau et al. (1993) used EI technology to insert full-length recombinant CD4 (rCD4), which has both the hydrophobic and extracellular domains, into the membrane of red blood cells (RBC). This dramatically increased the half-life of CD4, and the electroinserted rCD4 has been found to be immunologically active and stable in vivo. The RBC with the electroinserted rCD4, while circulating in the blood stream, can sponge off free HIV particles, and the aggregates formed by RBC/CD4 and RBC ought to be cleared rapidly by the reticuloendothelial system. RBC-CD4, where the rCD4 concentration in the RBC membrane is only 40 ng/mL, is able to block transmission of HIV from patient isolates to normal peripheral blood lymphocytes. Soluble CD4 at the higher concentration of 10 μg/mL, however, fails to block such transmission (Nicolau et al., 1993). Further clinical trials of this novel method of HIV treatment are planned.

Applications of Electroporation to Cardiology

Prevention of restenosis (reblockage of arteries following angioplasty procedure) is one of the greatest challenges facing the cardiologist community. Restenosis, which normally develops within 1–3 months post percutaneous transluminal coronary angioplasty (PTCA) costs the health care system \$1 billion/yr because the rate of repeat PTCA procedure is some 37%. Many different pharmacological approaches, new catheter designs, and clinical procedures have been tried, but none has proven successful.

Electropermeabilized platelets have been used as drug-delivery vehicles to prevent restenosis (El Gammal et al., 1992; Crawford, 1993 and references therein). They have shown that it is possible to inhibit platelet aggregation, even in the presence of a platelet activator. Autologous or homologous platelets are electroporated with an antiplatelet drug such as the prostcyclin analog Iloprost (MW: 361), reconstituted with RBC, then delivered at the site of the lesion. The in vivo work in mice shows an average reduction of 56% in platelet deposition related to restenosis. Experiments with the excised porcine aorta in a perfusion chamber, at different shear rates, shows similar results. Crawford is currently running extensive double blind studies with the pig model to reconfirm his observations by direct angiographic and morphometric measurements (personal communication).

Despite apparent success with the animal models, discussion with leading clinical cardiologists on such ex vivo treatment of platelets for prevention of restenosis has been discouraging. Ex vivo processing of blood components is viewed as problematic. We therefore think it would be more suitable to develop an in vivo electroporation catheter that can be used for angioplasty and the delivery of drugs/gene in situ. The procedure can serve as a prophylactic treatment for all angioplasty patients. Our calculations of total power and current requirements, based on typical dimensions of a coronary artery, show that such a catheter would be feasible and will not pose any undue risk. It will be suitable for delivery of drug/gene into media or adventia. In vivo electroporation may function better because of better internalization of the molecules.

GENE THERAPY: AN OVERVIEW

Gene therapy has progressed very rapidly to the point of limited clinical trials. Refer to a book and major reviews on the subject (Anderson, 1991; Larrick and Burck, 1991; Miller 1992; Mulligan 1993). Many vectors and delivery options are available, including retrovirus, adenovirus, herpes simplex virus, direct injection, and glycoprotein conjugate. Similarly, the number of cellular targets for gene therapy is also large. In the United States alone, as many as 52 protocols have been approved, including ADA (−) SCID, LDL receptor gene transfer for familial hypercholesteremia, advanced melanoma, and cystic fibrosis, to mention only a few.

Potter (1988) has discussed the application of electroporation to human gene therapy and has shown why this technique should be a viable alternative with some distinct advantages over the most predominantly used retrovirus approach. Keating and Toneguzzo (1990) have described a model for gene therapy using electroporation.

Since published results show that it is possible to transfer, by electroporation, a gene as large as 190 kb, we envisage all single gene diseases whose genomic size is less than this to be candidates for gene therapy. This would include all small genes such as globin and insulin and very large genes such as the 186-kb Factor VIII gene. If we take into consideration only the size of the cDNA, even the largest genes fit well within the electroporation regime. The cDNA for Duchenne Muscular Dystrophy (DMD) is only 16 kb, whereas the genomic size is greater than 2 Mb. Antisense therapy is now considered a new stable mate (Akhtar and Ivinson, 1993) of gene therapy. Since these are small, about 18 mer, no problem is anticipated in terms of the introduction of these oligos into cells, but there has been no known publication dealing with the stability problem of electroporated antisense oligos. Electroporation may also be useful for delivering aptamers (defined oligos) in vivo that bind to specific proteins.

Current Techniques: Advantages and Disadvantages

In this section we outline the advantages and disadvantages of electroporation versus retrovirus (RV), the current primary method of gene transfer. We also touch on the other methods. The advantages of electroporation, compared to retrovirus, are as follows: (1) it is applicable to all cell types, (2) it has the ability to transfer very large DNA, (3) it is not limited to replicating cells and can stably transfect progenitor as well as fully differentiated cells, (4) no helper cell lines are necessary, and (5) risk of insertional mutagenesis and unwanted transcription of cellular genes by viral LTR is greatly reduced.

The disadvantages of electroporation include current low efficiency of stable transfection, less than 0.1%, the possibility of damage to a portion of the target population leading to cell death, and the possible formation of concatatemers in the target genome.

The greatest advantage of the RV is the high efficiency of gene transfer and integration into target genome. Efficiency could, theoretically, reach 100% but in practice it is closer to 30%. RV also introduces a single copy of the gene on integration. Among the disadvantages of RV-mediated gene transfer is the possibility of cancer linked to such transfer. This was reported (Kolberg, 1992) for three rhesus monkeys with T cell lymphomas.

A few comments should be made about the current constraints with electroporation. Recent work on *Dictyostelium* (Kuspa and Loomis, 1992) shows a 20–60-fold increase in electroporation efficiency if the plasmid is cut with a restriction enzyme and the same enzyme is incorporated during electroporation. The same strategy can possibly be applied to mammalian cells. With the use of a square-wave pulser, the cell viability can go up more than 95% and Keating has shown that electroporation produces single or very low copy of the gene (Keating and Toneguzzo, 1990; Keating et al., 1990). Finally, it must be remembered that introduction of a large quantity of any DNA is potentially mutagenic.

RV-mediated gene therapy can only be applied to replicating cells. Recently, alternative means of gene transfer by adenovirus (AV) and herpesvirus (HV) are being tried out that do not have this problem. Davies (1992) has briefly summarized some of the recent research. Delivery of *E. coli* lacZ marker genes, incorporated in a replication-defective AV, into the rodent brain has also shown high expression, particularly in the first week after injection (Breakefield, 1993).

Electroporetic Gene Therapy: Clinical Applications

As mentioned before, we are not aware of any gene therapy clinical trial that has been based on electroporation. We feel we have provided enough rationale in the previous section, however, to conclude that the potential application of electroporetic gene therapy, both ex vivo and in vivo, remains extremely high. We would like to discuss a few applications that relate to both in vivo and ex vivo electroporation-mediated gene transfer.

Titomirov et al. (1991) have shown that it is possible to transfer genes in vivo in skin cells. This has been demonstrated by injection of pSV3neo and plasmid DNA subcutaneously into new born mice, followed by a short intense pulse (electrical field strength of 400–600 V/cm and a pulse width varying between 150 and 300 μsec) delivered across the epidermal and dermal tissue of the skin. Excision of the tissue and subsequent culture of skin fibroblast cells showed that stable transduction has taken place although the level of gene expression was low.

The first European electroporation-mediated gene therapy trial to treat cancer is planned by Mertelsmann in Germany (Abbott, 1992) and is based on in vivo activation of cytotoxic T cells. Mertelsmann has chosen for his clinical trial three types of cancers primarily because of their good response to IL-2, namely renal cell carcinoma, colon cancer, and malignant melanoma. The plan is to insert the gene by electroporation into autologous fibroblasts cultured from the skin biopsy of the patients. After stable transfection, these fibroblasts will be mixed with patients' cancer cells, irradiated, and reinjected subcutaneously.

Keating and Toneguzzo (1990) and Keating et al. (1990) have published major papers on the technique of electroporation for gene therapy. They have shown that EP is an efficient and reproducible method for gene transfer in permanent hematopoietic lines, genes are integrated in single or low copy number, and transferred genes are expressed at a frequency of more than 2%. By studying the effect of different viral and cellular promoters on the transient expression of reporter genes, Keating et al. (1990) have inferred that human marrow stromal cells may be an attractive gene delivery vehicle.

In the first human gene therapy study (personal communication), Keating's objective would be to confirm whether bone marrow stromal cells can be transplanted by using a selectable marker to identify the engrafted stromal population. The focus is on two clinical conditions, severe hemophilia B (FIX cDNA) and aplastic anemia.

An interesting example of gene therapy to cure hemophilia B has come to our notice. Dai et al. (1992) electroporated an RV vector containing FIX cDNA, driven by various promoters, into an amphotropic cell line and found that 75% of the secreted FIX was biologically active. The authors drew the conclusion that the FIX-deficient human primary skin fibroblast had been cured.

Strategies for Successful Gene Therapy using Electroporation

Despite the major advantages of electroporation technology that it would be free from viral vectors, would allow for insertion of whole genes, and may not require integration, it is natural to ask what are the areas in which progress must be made for electroporation to be successful in gene therapy. There are three areas: (1) improved delivery and prevention of cell death, (2) construction of DNA segments that can undergo homologous recombination (HR) by introducing sequences onto the DNA that drive HR (if such sequences exist), or (3) development of minichromosomes that would persist in the cell as episomes. These can be constructed from fragments of virus that do not undergo integration and have telomeric sequences at the end.

Ideally, the transfer of genes into eukaryotic cells should have the properties of high efficiency of transfer into target cells, efficient integration at the correct locus, normal expression and regulation, safety, and ease of handling and production. None of the currently available methods can be considered ideal, but a good start has been made and we do see electroporation playing a major role in the future.

INSTRUMENTATION FOR CLINICAL APPLICATIONS

A field that is as homogeneous as possible and of the correct amplitude is desirable because excessive field strength results in lysing of cells and a low field strength reduces efficacy. Tissue is generally a very inhomogeneous medium. Skin, especially the outer layer, the Stratum Corneum, exhibits a very high resistance at low voltages. This resistance diminishes by order of magnitudes if the top layers are removed. Furthermore, there are indications that this outer layer will experience a dielectrical breakdown at voltages typically used for electroporation, resulting in a dramatic reduction in resistivity (Reiley, 1992).

Electrical Field Generator Design

Electrical fields can be generated by applying a voltage to two electrodes in direct contact with the tissue or by induction from a fast varying magnetic field generated by a coil according to Faraday's law. The latter was found effective in arresting and reducing tumors in rats even without administering anticancer drugs (Costa and Hofmann, 1987).

In general, exponential decay or square wave forms are used to generate electrical fields in the direct contact mode. It appears that square wave forms generate less tissue damage than exponential wave forms. The two researchers who performed in vivo ECT selected square pulses as the preferred wave form (Okino et al. 1992; Mir et al. 1991a).

A desirable feature of a generator for in vivo work is the acceptance of an external feedback signal to set the charging voltage. In many applications, the applicator might be located in places where direct observation or measurement of the electrode gap is not possible. The electrode gap then needs to be automatically measured at the applicator, and an electrical signal indicating the gap needs to be relayed to the generator. A separate account will be published elsewhere giving details of the special requirements on the design of electrical field generators.

Applicators

Depending on the topological situation, several methods can be used to apply electrical fields to tumors. Outside electrodes (forceps or caliper types) can be used for small surface tumors or tumors accessible during operation; for example, open or laparoscopic procedures. Surface electrodes (e.g., with meander-type electrodes) are useful for large tumors on the body surface or large tumors accessible during open or laparoscopic procedures. Insertion electrodes (e.g., two parallel needles or an array of needles) are useful for deep-seated tumors. Liquid treatment of blood or other bodily fluid related cancers can be done in a batch mode or in an ex vivo flow through electroporation system with a repetitive pulsing generator.

SAFETY

The pulsed electrical fields that are needed for ECT are similar in amplitude to the fields that successfully electrotransform mammalian cells in vitro. They are in the range of a few 100 V/cm to several kilovolts per centimeter. These fields, although very short, are orders of magnitude higher than other fields associated with low-level therapeutic applications.

By localizing the fields to the tissue that needs to be treated, we can minimize disturbance of the body's critical electrical functions especially of the heart. Synchronizing the pulse with the heart beat is one possible method to minimize erratic effects of the fields (Okino et al., 1992).

Special precautions need to be taken to ensure safety of the patient in case of malfunction of the generator. The patient can be protected by selecting the magnetic parameters of the pulse transformer so that the transformer saturates if the pulse length exceeds 100 μsec by, for example, more than 40% and essentially cuts off the delivery of energy at this point. Another important safety feature of the instrument should be the complete isolation of the output circuit from any potential to which the patient might be connected.

FUTURE OUTLOOK

The examples given in this discussion clearly demonstrate that electroporation based drug/gene delivery has potential in this field. As is evident, however, there are challenges to be met. Conventional oncologists view electrochemotherapy as a "local" treatment, although preliminary data have been presented to show that there is an immune response. Also, the key role of aggressive local control of tumors is often overlooked despite the fact that a large percentage of cancer deaths can be avoided if such control can be exercised. The basic problem of ECT is that, until very recently, only two groups of researchers (Japan and France) have been actively involved in this modality of therapy.

A group in Sweden has started treating brain tumors using ECT (Salford et al., 1994) and two U.S. groups, the University of South Florida in collaboration with Genetronics Inc. and MIT/Harvard, have initiated experiments to reproduce Mir's results. BTX scientists have recently achieved highly significant results of tumor regression with human pancreatic and non-small cell lung cancer tumors transplanted subcutaneously on to nude mice which were treated with bleomycin and pulsed electric field. More active participation by different groups to use this enabling technology is sorely needed. It is hoped that this chapter will encourage the pursuit of this technology by additional researchers.

New applications of electroporation are emerging. Crawford has plans to coat stent with electroporated platelets to reduce restenosis. We also envisage that electroporation of existing antiproliferative and anticoagulative drugs like heparin in monocytes or neutrophils can aid in this process. In vivo targeting of inflamed areas by electroporated neutrophils has already been carried out (Sixou and Teissie, 1992). Mangal and Kaur (1991) have also shown that the rate of elimination of a drug electroloaded in erythrocytes was considerably reduced compared to the free drug pointing to the possibility of sustained release by such delivery method. Electroinsertion technology has many potential applications other than the reported encapsulation of inositol hexaphosphate in RBC to improve oxygenation (Mouneimne et al., 1990).

The possibility of delivering nerve growth factor (NGF) in the brain following stereotactic implantation for treatment of Parkinson's disease is perfectly feasible, as is the direct in vivo gene transfer into myofibers for application in DMD. In all these cases, the challenges are not just the optimization of electrical parameters and the use of appropriate drugs, but also the generation of appropriate electrical fields, since these will invariably involve tissue electroporation.

Although in the short span of a decade or so, the science of electroporation has become a standard laboratory technique for introducing exogenous molecules into cells, the reported range of applications has been enormous, varying from antisense expression on human leukemia cell proliferation, role of TNF-α in pathogenesis, regulation of transcription in cardiac muscle and HIV-1 expression in the fetus, to name only a few.

ACKNOWLEDGMENTS

SBD thanks Joe Snodgrass and Dr. Milton Taylor for many useful discussions and Dr. Jim Weaver for sending papers before publication.

References

Abbott, A. (1992). German state unexpectedly approves first gene trials. Nature. **360**: 702.

Abraham, W., Bommannan, D. M., Potts, R. W., and Tamada, J. (1993). Drug Delivery: Status, Issues and Challenges for Medicine. Page 6. In Proc. BES Symp., Bielefeld, April 5–8.

Akhtar, S., and Ivinson, A. J. (1993). Therapies that make sense. Nature Genetics. **4**: 215–216.

Anderson, W. F. ed., (1991). Human Gene Therapy. **2** (2).

Belehradek, J. Jr., Orlowski, S., Poddevin, B., et al. (1991). Electrochemotherapy of spontaneous mammary tumors in mice, Eur. J. Cancer. **27**: 73–76.

Breakefield, X. W. (1993). Gene delivery into the brain using virus vectors. Nature Genet. **3**: 187–189.

Costa, J. L., and Hofmann, G. A. (1987). Malignancy Treatment. US Patent, 4,665,898.

Crawford, N. (1993). Electropermeabilized platelets as drug delivery vehicles. Pages 284–294. In Restenosis Summit V, Cleveland Clinic Foundation.

Dai, Y. F., Qui, X. F., Xue, J. L., and Liu, Z. D. (1992). High efficient transfer and expression of human clotting Factor IX cDNA in cultured human primary skin fibroblasts from hemophilia B patient by retroviral vectors; Science in China. Ser. B. **35**: 183.

Davies, K. (1992). Moving straight to the target. Nature. **358**: 519.

El Gammal, B. A. B., Pfliegler, G., and Crawford, N. (1992). Effect of platelet encapsulated Iloprost on platelet aggregation and adhesion to collagen and injured blood vessels in vitro. Thrombosis and Haemostasis. **68**: 606–624.

Gregoriadis, G., and Florence, A. T. (1993). Liposomes in drug delivery: Clinical, Diagnostic, and Ophthalmic Potential. Drugs. **45**: 15–28.

Hofmann, G. A., and Dev, S. B. (1993). Electroporation: From Research Laboratories to Clinical Practice. Proceedings of IEEE Engineering in Medicine and Biology. **15**: 11420–1421.

Kanesada, H. (1990). Anticancer effect of high voltage pulses combined with concentration dependent anticancer drugs on Lewis lung carcinoma *in vivo* [in Japanese, English summary and tables]. J. Jpn. Soc. Cancer Ther. **25**: 2640–2648.

Keating, A., Horsfall, W., Hawley, R. G., and Toneguzzo, F. (1990). Effect of different promoters on expression of genes introduced into hematophoietic and marrow stem cells by electroporation. Expl. Hematol. **18**: 99–102.

Keating, A., and Toneguzzo, F. (1990). Gene transfer by electroporation: A model for gene therapy. Pages 491–498. In Progress in Clinical and Biological Research: Bone Marrow Purging and Processing. Gross, S., et al. eds. Alan R. Liss, New York.

Kolberg, R. (1992). Gene-transfer virus contaminant linked to Monkey's cancer. J. NIH Research. **4**: 43–44.

Kuspa, A., and Loomis, W. F. (1992). Tagging developmental genes in *Dictyostelium* by restriction-mediated integration of plasmid DNA. Proc. Natl. Acad. Sci. USA. **89**: 8803–8807.

Langer, R. (1990). New Methods of Drug Delivery. Science. **249**: 1527–1533.

Larrick, J. W., and Burck, K. L. (1991). Gene Therapy, Elsevier, New York.

Mangal, P. C., and Kaur, A. (1991). Electroporation of red blood cell membrane and its use as a drug carrier system. Indian J. Biochem. Biophys. **28**: 219–221.

Miller, A. D. (1992). Human Gene therapy comes of age. Nature. **357**: 455–460.

Mir, L. M., Banoun, H., and Paoletti, C. (1988). Introduction of definite amounts of nonpermeant molecules into living cells after electropermeabilization: Direct access to cytosol. Exptl. Cell Res. **175**: 15–25.

Mir, L. M., Belehradek, M., Domenge, C., et al. (1991a). Electrochemotherapy, a novel antitumor treatment: first clinical trial. C. R. Acad. Sci. Paris, **313**: 613–618.

Mir, L. M., Orlowski, S., Belehradek, J. Jr., and Paoletti, C. (1991b). Electrochemotherapy potentiation of antitumor effect of bleomycin by local electric pulses. Eur. J. Cancer. **27**: 68–72.

Mir, L. M., Orlowski, S., Poddevin, B., et al. (1992). Electrochemotherapy tumor treatment is improved by interleukin-2 stimulation of the hosts's defenses. Eur. Cytokine Netw. **3**: 331–334.

Mouneimne, Y., Barhoumi, R., Myers, T. et al. (1990). Stable rightward shifts of the oxyhemoglobin dissociation curve induced by encapsulation of inositol hexaphosphate in red blood cells using electroporation. FEBS Lett. **275**: 117–120.

Mouneimne, Y., Tosi, P.-F., Barhoumi, R. and Nicolau, C. (1992). Electroinsertion: An electrical method for protein implantation in cell membranes. Pages 327–346. In Guide to electroporation and electrofusion. Chang, D. C., Chassy, B. M., Saunders, J. A., and Sowers, A. E., eds. Academic, San Diego.

Mulligan, R. C. (1993). The basic science of gene therapy, Science. **260**: 926–931.

Nicolau, C., Volsky, D. J., and Potash, M. J. et al. (1993). Electroinserted full length CD4 as an active viral receptor on the plasma membrane of erythrocytes. Page 4. In Proc. BES Symp., Bielefeld, April 5–8.

Okino, M., and Mohri, H. (1993). Effects of a high-voltage electrical impulse and an anticancer drug on in vivo growing tumors. Jpn. J. Cancer Res. (Gann) **78**: 1319–1321.

Okino, M., Tomie, H., Kanesada, H., and Marumoto, M., et al. (1992). Optimal electric conditions in electrical impulse chemotherapy. Jpn. J. Cancer Res. **83**: 1095–1101.

Potter, H. (1988). Electroporation in Biology: Methods, Applications, and Instrumentation. Anal. Biochem. **174**: 361–373.

Reiley, J. P. (1992). Electrical simulation and electrophysiology. Page 31. Cambridge University Press, New York.

Salford, L. G., Persson, R. B. R., and Brun, A. et al. (1994). A brain tumor therapy combining bleomycin with ion vivo electropermeabilization. Biochem. Biophys. Res. Commun. **194**: 938–943.

Sixou, S., and Teissie, I. (1992). In vivo targeting of inflammed areas by electroporated neutrophils. Biochem. Biophys. Res. Commun. **186**: 860–866.

Titomirov, A. V., Sukharev, S., and Kistanova, E. (1991). In vivo electroporation and stable transformation of skin cells of new born mice by plasmid DNA. Biochim. Biophys. Acta. **1088**: 131–

Wallace, B. M., and Lasker, J. S. (1993). Stand and deliver: Getting peptide drugs into the body. Science. **260**: 912–913.

CHAPTER 10

Electroporation and Transgenic Plant Production

M. Joersbo

J. Brunstedt

ABSTRACT

Electroporation is a well-established method for production of trans-genic plants. Short high-voltage pulses can permeabilize the proto-plast plasma membrane, facilitating uptake of plasmid DNA that can become expressed transiently and, eventually, be stably incorporated into the genome.

The major electrical parameters are field strength and pulse dura-tion, which are inversely related and can be chosen within wide ranges (100–5000 V / cm and 0.01–100 msec). Stable transformation requires less rigorous electrical conditions than transient expression. Transient and stable transformation increase with plasmid DNA con-centration, up to about 100 μg / mL; addition of carrier DNA lowers the amount of plasmid DNA required for transformation. Linearized plasmid DNA and heat shock enhance stable transformation. Addition of PEG stimulates transient expression and, in most cases, stable transformation. The transformation rate is also affected by protoplast size, pulse type, culture medium, and temperature.

Stable transformation frequencies are in the range 0.0001–0.1% of the electroporated protoplasts. Transgenic plants contain, on average, from one to three copies of the exogenous gene, and all copies are usually integrated into one site in the genome. The inserted plasmid DNA is often modified by rearrangement and ligation events, and the copy number does generally not correlate with expression level. Transgenic plants regenerated from electroporated protoplasts are most often fertile, and the exogenous genes appear to be inherited as a single dominant character in a Mendelian fashion.

Although the cell wall is generally regarded to be impermeable to DNA, some intact cells and tissues can be induced to take up DNA by electroporation.

INTRODUCTION

Electroporation has proved to be a successful method for gene transfer into plant protoplasts. The large number of plant species transformed suggests that electroporation can stimulate DNA uptake in most, if not all, plant species. This apparent lack of limitations has made electroporation particularly useful for the transformation of cereals.

Electroporation is a simple technique. It requires a device for delivery of high voltage pulses, but since the electrical parameters can be varied within wide limits, it should only meet a few requirements. Electroporation can even be performed with alternating current directly from the mains (Joersbo and Brunstedt, 1990a).

Expression of genes transferred to plant protoplasts by electroporation was first reported by Fromm et al. (1985). They subjected carrot, maize, and tobacco protoplasts to short, high-voltage pulses in the presence of plasmid and measured the level of expression of a marker gene following DNA uptake. Since then, transient expression has been obtained in protoplasts from more than 25 species (Table 10-1).

By some largely unknown mechanism, exogenous DNA is incorporated into the genome, leading to stably transformed protoplasts which, in turn, can be regenerated to transgenic calli and eventually to plants. Tobacco was the first plant to be transformed by electroporation (Shillito et al., 1985; Riggs and Bates, 1986); it was followed by several others, including the commercially important cereals maize and rice.

In most cases, the transgenic plants display a normal phenotype and are fertile. Transmission of the integrated gene to the offspring is often, as would be expected for a Mendelian inheritance (Table 10-2). Protoplasts of some species are, however, difficult to regenerate through to fertile plants. This has encouraged studies of the electroporation of intact cells and tissues that from some species have been shown to be amenable to electrotransformation (e.g., Lindsey and Jones, 1987a).

TABLE 10-1. Transformation of Plant Protoplasts[a]

Species	TE	STC	TP	References
Alnus incana	+	−	−	Seguin and Lalonde (1988)
Beta vulgaris (sugar beet)	+	+	−	Lindsey and Jones (1987b, 1989); Joersbo and Brunstedt (1990a)
Brassica napus (rape seed)	?	+	+	Guerche et al. (1987b)
Dactylis glomerata (orchard grass)	?	+	+	Horn et al. (1988)
Daucus carota (carrot)	+	−	−	Fromm et al. (1985) Boston et al. (1987)
Glycine max (soybean)	+	+	+	Hauptmann et al. (1987) Christou et al. (1987) Widholm et al. (1992)
Helianthus annuus (sunflower)	+	−	−	Kirches et al. (1991)
Lactuca sativa (lettuce)	+	+	+	Chupeau et al. (1989)
Larix eurolepis (larch)	+	−	−	Charest et al. (1991)
Lycopersicon esculentum (tomato)	+	−	−	Tsukada et al. (1989)
Lycopersicon peruvianum	+	+	+	Bellini et al. (1989)
Nicotiana tabacum (tobacco)	+	+	+	Fromm et al. (1985) Shillito et al. (1985) Riggs and Bates (1986)
Oryza sativa (rice)	+	+	+	Ou-Lee et al. (1986) Toriyana et al. (1988) Zang et al. (1988) Shimamoto et al. (1989)
Panicum maximum	+	−	−	Hauptmann et al. (1987)
Pennisetum purpureum (napier grass)	+	−	−	Hauptmann et al. (1987)
Persea americana (avocado)	+	−	−	Percival et al. (1991)
Petunia hybrida	+	+	+	Hauptmann et al. (1987) Tagu et al. (1990)
Picea glauca (white spruce)	+	−	−	Bekkaoui et al. (1988)
Picea mariana (black spruce)	+	−	−	Tautorus et al. (1989)

(Continued)

TABLE 10-1. (*Continued*)

Species	TE	STC	TP	References
Pinus banksiana (jack pine)	+	−	−	Tautorus et al. (1989)
Pisum sativum (pea)	+	+	−	Hobbs et al. (1990) Puonti-Kaerlas et al. (1992)
Saccharum spp. (sugarcane)	+	+	−	Hauptmann et al. (1987) Rathus and Birch (1992a, 1992b)
Solanum brevidens (wild potato)	+	−	−	Jones et al. (1989)
Solanum tuberosum (potato)	+	−	−	Jones et al. (1989)
Sorghum bicolor (sorghum)	+	−	−	Ou-Lee et al. (1986)
Triticum monococcum	+	−	−	Ou-Lee et al. (1986)
Triticum aestivum (wheat)	+	−	−	Oard et al. (1989)
Vigna aconiti folia (moth bean)	+	+	+	Köehler et al. (1987a, 1987b)
Zea mays (maize)	+	+	+	Fromm et al. (1985) Fromm et al. (1986) Rhodes et al. (1988)

[a]List of plant species transformed by electroporation. It is indicated, for each species, whether transient expression (TE), stably transformed callus (STC), and/or transgenic plants (TP) has been obtained.

ELECTRICAL PARAMETERS

When a protoplast is exposed to an electrical pulse, it can remain unaffected, be reversibly permeabilized (allowing DNA uptake), or be ruptured (for a review, see Joersbo and Brunstedt, 1991). The impact of the pulse depends on electrical conditions, protoplast size, and membrane stability and fluidity.

Field Strength

The field strength is the most important electrical factor. Dose–response curves typically show that low field strengths result in neglible levels of transient expression (Fig. 10-1a). This is presumably related to the phe-

TABLE 10-2. Inheritance of Aminoglycoside Resistance[a]

Target Plant	Plant no.	Percentage of resistant progeny	Reference
Lettuce	AKr25, AKr26 AKr34, AKr53	78%, 55% 76%, 52%	Chupeau et al. (1989)
Petunia	1a, 5b	73%, 74%	Tagu et al. (1990)
Rape seed	PG20	73%	Guerche et al. (1987b)
Rice	3-1, 6-1 6-3, 8-1	95%, 89% 86%, 69%	Shimamoto et al. (1989)
Rice	S-1, S-7	85%, 73%	Tada et al. (1990)
Rice	—	75%	Battraw and Hall (1992)
Tobacco	101, 103, 108 109, 120, 121 123, 143	76%, 77%, 77% 74%, 75%, 68% 69%, 71%	Riggs and Bates (1986)

[a]Transgenic plants regenerated from electroporated protoplasts were self-fertilized and the offspring tested for aminoglycoside resistance. For a single dominant trait inherited in a Mendelian fashion, 75% resistant offspring would be expected.

nomenon of critical voltage, which has been described for artificial membranes (Zimmermann et al., 1974; Coster and Zimmermann, 1975). According to this, the voltage drop over the protoplast must exceed a certain threshold before electropermeabilization occurs. Above this field strength, transient expression increases rapidly and approximately linearly with increasing field strength, until it levels off at a field strength where a significant number (often 50–70%) of the protoplasts has been killed by the electrical pulse. Further increase of field strength leads to decreasing transient expression, as well as protoplast viability.

Transient expression in tobacco mesophyll protoplasts was very low when the latter were electroporated at field strengths up to 200 V/cm. Increasing the field strength to 300 V/cm resulted in a 20–30-fold increase in gene expression (Guerche et al., 1987a). Similarly, in carrot protoplasts, a twofold increase in field strength (200–400 V/mm) resulted in eightfold more transient expression (Hauptmann et al., 1987).

Pulse Duration

The duration of an electrical pulse affects electroporation efficiency less than the field strength. In sugar cane protoplasts, a doubling of the pulse

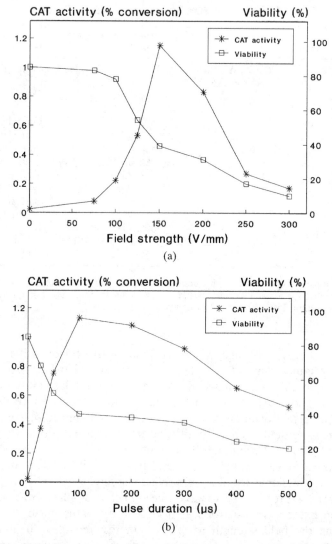

Figure 10-1. Typical dose–response curves for electrical field strength and pulse duration. Sugar beet protoplasts were electroporated with a chloramphenicol acetyltransferase (*cat*) gene, and transient expression (*cat* activity) and viability were measured. Electroporation with square pulses at (a) increasing field strengths for 100 μsec and (b) increasing pulse durations at 150 V/mm (Joersbo, 1990).

duration from 5 to 10 msec resulted in 40% higher transient expression (Rathus and Birch, 1992a). In sugar beet protoplasts, transient expression increases approximately linearly with pulse duration until a plateau is reached, after which it declines gradually (Fig. 10-1b). No measurable minimum or critical pulse length has been found. Benz and Zimmermann (1980) estimated the molecular rearrangements leading to formation of pore structures to take about 10 nsec.

Number of Pulses

When a series of consecutive pulses is delivered, the effect of each pulse on electroporation efficiency appears to be additive. Guerche et al. (1987a) found, in tobacco mesophyll protoplasts, that transient expression increased in a linear fashion with the number of pulses, up to 7 pulses. More than 10 consecutive pulses reduced transient expression, probably due to reduced protoplast viability. Similar results were obtained by Ou-Lee et al. (1986), who pulsed wheat protoplasts with up to 80 pulses.

Quantitative Relationship

The relative importance of the two parameters, field strength and pulse duration, has been described for rectangular pulses using dye uptake as a marker for electropermeabilization (Joersbo and Burnstedt, 1990b). The fraction of stained protoplasts, p, after electroporation at the field strength, E (V/mm), for the duration, t (μsec) can be fitted to the equation

$$p = 1 - (1/2)^{t^a E^b / c} \tag{1}$$

where a is 1.0–1.4, b is 2.7–2.9, and c is 345–1653 $\times$ 10^6, depending on the species. c was found to be roughly proportional to the square of the protoplast diameter.

Solving Equation 1 for 50% stained protoplasts gives

$$t^a E^b = c \tag{2}$$

demonstrating quantitatively the inverse relationship between field strength and pulse duration, the field strength having a considerably higher impact than the pulse duration. Thus, for example, for carrot protoplasts, a 150

V/mm pulse for 800 μsec is equally efficient as a 300 V/mm pulse for only 115 μsec.

Protoplast Size

The voltage drop over a protoplast is critical in determining whether or not electropermeabilization occurs. As field strength is equal to voltage drop per distance unit (usually millimeters or centimeters), large protoplasts will, at a given field strength, experience a higher voltage drop than smaller, but otherwise identical, protoplasts. As a result, large protoplasts are permeabilized at low field strengths.

Zimmermann and Benz (1980) studied individual protoplasts of guard cells of *Vicia faba* and found that the required field strength for electropermeabilization decreased with increasing diameter (all protoplasts required the same voltage drop, irrespective of their diameter). Likewise, Rouan et al. (1991) found an approximately inverse linear relationship between protoplast diameter and the field strength required for electropermeabilization.

Electrical Pulse Types

The most commonly used pulses are produced by discharging a capacitor, resulting in pulses with exponentially decaying voltage. Other types are also used, including rectangular and AC pulses.

Exponentially Decaying Pulses

The widespread use of exponentially decaying pulses is probably related to the fact that the devices for production of such pulses are relatively easily constructed. They consist of a bank of capacitors (10–1000 μF) and a DC power supply for charging the capacitors (to about 100–500 V). Pulse durations are usually in the range 1–50 ms and depend on capacitance, conductivity, temperature, and volume of medium in the electroporation chamber. Therefore, the pulse duration should be monitored carefully by an oscilloscope, otherwise comparisons are difficult between different electrical treatments and experimental setups.

Rectangular Pulses

Using rectangular pulses, the voltage is constant throughout the pulse period. They offer the advantage of being independent of medium conductivity and volume. Compared to exponentially decaying pulses, the voltage of rectangular pulses is generally higher (100–2000 V) and the pulse duration considerably shorter (10–1000 μsec) (Potter, 1988).

Alternating Current Pulses

In the absence of an electroporator, alternating current directly from the mains (220 V) can be used. The pulse duration is controlled by the blowout of a small fuse inserted between the ends of an electrical cord and a suitable electroporation chamber (Joersbo and Brunstedt, 1990a). Depending on the fuse tolerance (50–400 mA), the pulse duration can be selected in the range of 0.5–10 msec. This is a short period of time compared to the 20-msec period of ordinary 50 Hz AC. The melting integral (m) determining fuse blowout is, however, given by $m = tI^2 = tU^2/R^2$, which in turn resembles Equation (2) (see earlier), relating field strength and pulse duration to electroporation efficiency. Thus, if the voltage is low when switching on, the time taken before fuse melting occurs will be accordingly longer and vice versa.

Comparison of Pulse Types

The various pulse types appear to affect individual species differently. For sugar beet protoplasts, alternating current and exponentially decaying pulses seem to be superior to rectangular pulses. Tobacco leaf protoplasts responded differently, the exponentially decaying and rectangular pulses giving highest transient expression (Joersbo and Brunstedt, 1990a).

Electroporation Medium

The electroporation medium serves the dual purpose of conducting electrical current and sustaining protoplast viability during the electrical shock. The electrical conductivity is particularly important when using capacitor discharge pulses, as the pulse length decreases with increasing conductivity of the medium.

Several media have been used. Both high-conductivity phosphate-buffered saline (Fromm et al., 1985) and low-conductivity CPW salts (Frearson et al., 1973; Seguin and Lalonde 1988) solutions have been used, but it is not clear which type is preferable.

Ca^{2+} ions are usually added to the electroporation medium, because of their membrane-stabilizing properties. About 4–5-mM Ca^{2+} ions significantly enhanced transient expression in carrot protoplasts, compared to 0- or 10-mM Ca^{2+} ions (Fromm et al., 1985). The presence of 5-mM Mg^{2+} ions resulted in only half of the level of transient expression as obtained with 5-mM Ca^{2+} ions in tobacco mesophyll protoplasts; Zn^{2+} ions almost completely inhibited transient expression (Guerche et al., 1987a). Elimination of chloride ions has been shown to increase cell survival and subsequent transformation efficiency, probably because chloride ions can be converted to toxic levels of Cl_2 by electrolysis (Tada et al., 1990).

The pH of the medium affects the required electrical potential. In the alga *Chara australis*, electropermeabilization was found at 0.25 V at pH 4.5, whereas at pH 9.0 it occurred at 0.45 V. This may be related to the degree of ionization of surface compounds on the plasma membrane that are considered to be involved in pore formation (Coster, 1968).

Electroporation Temperature

The temperature at which electroporation is performed significantly affects both the fluidity of the plasma membrane and the conductivity of the medium. Studying the alga *Valonia utricularis*, Zimmermann and Benz (1980) found that the field strength needed for electropermeabilization increased more than two-fold (1.6–3.6 V) when the temperature was decreased from 25 to 3°C. Many plant protoplast electroporation protocols prescribe to electroporate on ice. Additionally, after electroporation, protoplasts are often incubated on ice. This prolongs the duration of the high permeability state of the electroporated protoplasts (Lindsey and Jones, 1987a), which, in turn, may enhance DNA uptake.

TRANSIENT EXPRESSION

Transient expression has been obtained in electroporated protoplast of several species (Table 10-1). It is particularly useful for rapid establishment of gene transfer protocols for stable transformation. If the electrical conditions are chosen so that about half of the protoplasts survive the electrical shock, the treatment is most likely to result in some DNA uptake and subsequent expression.

Expression of genes introduced by electroporation is detectable after about 12 hr, reaches a maximum after 1–2 days, then declines gradually to a very low level during the following 6–10 days (Hauptmann et al., 1987, Seguin and Lalonde, 1988). In addition to the electrical conditions and time after electroporation, transient expression levels strongly depend on the concentration of DNA in the electroporation medium.

Plasmid DNA Concentration

Transient expression increases with increasing plasmid DNA concentration up to about 100–200 $\mu g/mL$, when saturation is reached. A linear correlation between transient expression and plasmid DNA concentration up to 40 μg/mL has been found in electroporated carrot protoplasts (Fromm et al., 1985) and up to 80 μg/mL in protoplasts of *Alnus incana* (Seguin and Lalonde, 1988). Interestingly, electroporated sugar cane protoplasts showed no expression at concentrations below 25 μg/mL, and as much as 250 μg/mL was required for maximum transient expression (Rathus and Birch 1992a).

Several plants have nucleases in the plasma membrane or cytosol (Wilson, 1968a, 1968b; Sawicka, 1987). Protoplasts rich in nucleases presumably require higher plasmid concentration at electroporation. Sugar beet protoplasts were able to degrade all added plasmid DNA after 2 min of incubation before electroporation, whereas tobacco protoplasts did not degrade any after 10 min. The saturating plasmid DNA concentrations were 100 μg/mL and 30 μg/mL for sugar beet and tobacco protoplasts, respectively (Joersbo, 1990).

Carrier DNA

Sonicated DNA from calf thymus or herring sperm is often added as carrier to reduce DNAse degradation of plasmid DNA. Carrier DNA reduces the amount of plasmid required for transformation. Guerche et al. (1987a) reported that 5 μg/mL of plasmid and 50 μg/mL of carrier DNA results in the same level of transient expression as 30 μg/mL plasmid without carrier DNA. Tautorus et al. (1989) found a 27% stimulation of transient expression in jack pine protoplasts following addition of 50 μg/mL sonicated calf thymus DNA. At 150 μg/mL, the carrier DNA had no effect. It can be speculated that such high carrier DNA concentrations can restrict uptake of plasmid DNA due to a limited number of sites available for penetration of DNA into electropermeabilized protoplasts.

Configuration of Plasmid DNA

Closed circular supercoiled plasmid has been used in most studies, but linearized plasmids have also been used. In black spruce protoplasts supercoiled plasmid was superior, giving 35% higher transient expression than linearized plasmid DNA, whereas in jack pine protoplasts linearized plasmid resulted in 2.5-fold higher expression than supercoiled (Tautorus et al., 1989).

Heat Shock and Polyethylene Glycol (PEG)

Compared with stable transformation (see below), heat shock does not seen to enhance transient expression. In carrot (Boston et al., 1987) and

Alnus incana (Sequin and Lalonde, 1988) protoplasts, no detectable effect of heat treatment at 45°C for 5 min before electroporation was found on transient expression. PEG facilitates DNA uptake into plant protoplasts (Krens et al., 1982), and a combination of PEG and electroporation can result in enhanced levels of transient expression. For example, transient expression was increased more than threefold in carrot (Boston et al., 1987) and *Alnus incana* (Seguin and Lalonde, 1988) protoplasts with a PEG-electroporation procedure.

TRANSGENIC CALLI AND PLANTS

Several plants have been stably transformed by electroporation (Table 10-1). Stable transformants can be produced by electroporating at the same electrical conditions as used for transient expression. It appears, however, that the maximum number of stable transformants is found at more gentle electrical conditions, which increase survival and, consequently, the number of transformants.

In tobacco protoplasts, maximum transient expression was found after eight pulses (300 V/cm, 16 μF capacitor), whereas the maximum number of stable transformants was obtained using five pulses of the same strength. The plating efficiency was reduced 53% by eight pulses and 35% by five pulses (Guerche et al., 1987a). Similarly, with sugarcane protoplasts where nine pulses (385 V/cm, 10 msec) were optimal for transient expression, from five to seven pulses resulted in the highest frequency of stably transformed calli (Rathus and Birch, 1992a, 1992b). Generally, a protoplast survival of 50–70% seems to be optimal for stable transformation, whereas maximum transient expression is often found when only 30–50% of the treated protoplasts remain viable.

Plasmid DNA

The optimal plasmid DNA concentration for stable transformation is similar to that used for transient expression. The transformation efficiency of sorghum protoplasts increased linearily with plasmid DNA concentration from 3–50 μg/mL (Battraw and Hall, 1991), and for tobacco protoplasts 50–150 μg/mL resulted in highest transformation frequencies (Shillito et al., 1985).

Carrier DNA apparently does not increase stable transformation rates (Battraw and Hall, 1991), but it can reduce the amount of plasmid required (Shillito et al. 1985). Both linearized and supercoiled plasmid can be used for stable transformation. Linearized DNA has been reported to be superior, resulting in 2.5–10-fold more tobacco transformants than with supercoiled DNA (Shillito et al., 1985). Likewise, in electroporated sorghum protoplasts, linearized plasmid DNA doubled transformation efficiency (Battraw and Hall, 1991).

Heat Shock and PEG

Heat shock before electroporation can increase plating efficiency considerably, resulting in 14-fold higher stable transformation rates in rice (Zhang et al., 1988). Tyagi et al. (1989) obtained enhanced transformation frequencies using a combination of electroporation and PEG. They observed that maximum transformation rates of tobacco protoplasts were found at a lower field strength in the presence of PEG (1 kV/cm) compared to 2 kV/cm in the absence of PEG. In contrast, Köehler et al. (1987b) reported 10–40% less transformed tobacco and moth bean protoplast-derived colonies when using PEG in combination with electroporation. Using a combination of electroporation, PEG, and heat shock Shillito et al. (1985) obtained very high transformation frequencies of tobacco protoplasts, in the order of 1–2% of the protoplast populations.

Transformation Frequency

Transformation frequencies obtained by electroporation are in the range of 1.0×10^{-6} to 1.0×10^{-3} with a typical average of 10×10^{-4}, calculated as the number of transgenic calli divided by the initial number of electroporated protoplasts. Examples of transformation frequencies are lettuce 1.0×10^{-3} (Chupeau et al., 1989), orchard grass 1.0×10^{-4} (Horn et al., 1988), rice 4.0×10^{-6} (Toriyama et al., 1988), and soybean 1.0×10^{-6} (Christou et al. 1987).

In addition to being dependent on the plant species, transformation frequencies also vary significantly between different varieties of the same species. Thus, using different cultivars of moth bean, Köehler et al. (1987b) obtained transformation frequencies from zero to 1.2×10^{-5}. With protoplasts of different tobacco cultivars, Tyagi et al. (1989) reported transformation frequencies ranging from 1.0×10^{-6} to 1.0×10^{-3}. Transformation frequency can also be affected by the ability of the protoplasts to express the introduced gene. Bower and Birch (1990) electroporated carrot protoplasts and found that no more than 20–25% of the protoplasts were capable of expressing a reporter gene, irrespective of the plasmid DNA concentration in the electroporation medium. This suggests that only a fraction of a protoplast population is competent to be transformed.

Integration Mechanism

Integration of exogenous DNA into the recipient genome is believed to be a random process, although homologous recombination has been suggested to be involved. Wirtz et al. (1987) co-transformed tobacco protoplasts with two vectors each containing a different and inactive fragment of a *neo* gene. Recovery of transformed kanamycin-resistant colonies indicated that homologous recombination or gene conversion had occurred in such a way that a functional *neo* gene had been created. Lurquin and Paszty (1988) were, however, unable to demonstrate homologous recombination in tobacco protoplasts electroporated in the presence of plasmids containing a *neo* gene and, in some experiments, a fragment of a 1,5-ribulose bisphosphate carboxylase (Rubisco) gene. The transformation frequency was not enhanced by the presence of the Rubisco fragment.

Homologous recombination would be expected to reduce the variability between independent transformants. Dean et al. (1988) used the same approach as Lurquin and Paszty (1988), but they found that the presence of a Rubisco sequence did not reduce the large differences in gene expression levels between transformants.

Copy Number

One or more copies of the foreign gene can be inserted into the genome of electroporated protoplasts. Less than half of transgenic plants contain only a single intact copy; the majority contains a few or several copies (Battraw and Hall, 1992). The average integration number is from two to three copies per genome, but numbers as high as 20 have been reported (Huang and Dennis, 1989).

It appears that the copy number varies within these limits for all transgenic species produced by electroporation. It is not known whether controllable parameters such as field strength and plasmid concentration in the electroporation medium can affect the copy number.

Integration Sites

The number of integration sites can be assessed indirectly by analysis of the inheritance pattern. With exceptions, introduced antibiotic-resistance genes are transmitted to seed progeny in a resistant/sensitive ratio of 3 : 1 upon self-fertilization (Riggs and Bates, 1986; Guerche et al., 1987b; Chupeau et al., 1989; Battraw and Hall, 1992). This ratio suggests that all copies are inserted into only one site in the genome. A 3 : 1 ratio does not, however, exclude multiple integrations if some of the copies are silent or if they are so closely linked that they are inherited as one gene.

Integration of two intact genes at unlinked sites would result in 87.5% resistant offspring and the unlinked genes in 93.8%. Such high percentages have been observed (Table 10-2). suggesting that multiple independent integrations may occur.

Integration Pattern

The integrated DNA is often modified considerably. In most of the plant species electrotransformed, more than half of the transformants show complex patterns by Southern analysis (Fig. 10-2). Some of these patterns demonstrate the occurrence of integrated concatemers, where two plasmids are ligated either head to head or head to tail. Concatemers have been found following electroporation with both linearized (Riggs and Bates, 1986) and supercoiled plasmids (Huang and Dennis, 1989). Ligation

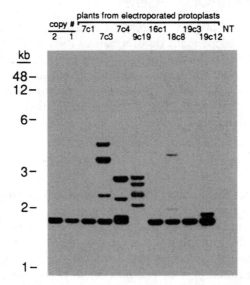

Figure 10-2. Southern blot of DNA from eight rice plants regenerated from protoplasts electroporated with pNEOGUS15. The expected 1.7-kb fragment is seen in seven lanes, and more or less complex restriction patterns are found in five lanes. Rice DNA was digested with EcoRI and HindIII. Control lanes include digested pNEOGUS15 DNA equivalent to one or two copies per genome or DNA from non-transformed (NT) rice (Battraw and Hall, 1992; with permission from Dr. T. C. Hall).

events have been implicated in up to 80% of the transformants of maize (Huang and Dennis, 1989). In many cases, the complex restriction patterns are not fully understood. One reason may be the loss of restriction sites. Modification of the free ends of linearized plasmids has been suggested to occur before or during integration, making the DNA resistant to restriction enzyme digestion (Riggs and Bates 1986). Loss of sequences not critical to gene expression can also occur. In transgenic sorghum calli, all of the resistant calli containing only partial genes expressed the inherent aminoglycoside-resistance gene (Battraw and Hall, 1991). When transforming with two genes on one plasmid, both genes are generally inserted. Battraw and Hall (1991) electroporated sorghum protoplasts with a plasmid containing both a kanamycin-resistance gene and a beta-glucuronidase (*gus*) gene and found that 63% of the kanamycin resistant calli also contained the expected *gus* sequence. When the two genes were present on two different plasmids, 38–42% of the resistant transformants contained both genes.

Expression of Introduced Genes

The expression level of genes introduced by electroporation typically varies 10-fold between independent transformants (Tada et al., 1990). The number of inserted copies of the foreign gene might be anticipated to affect the expression level. Although a correlation between copy number and expression level has been reported in electrotransformed tobacco calli/plants (Shillito et al., 1985), such a correlation is absent in most studies (Battraw and Hall, 1992).

Fertility

Transgenic plants regenerated from electroporated protoplasts are generally fertile. In a few of the early electroporation papers, however, there was no mention of the fertility of transgenic plants. Presumably, this could be related to the protoplast source rather than to the electroporation procedure, since plants regenerated from protoplasts isolated from cells cultured for extended periods can lack fertility. Battraw and Hall (1992) found that embryogenic rice cell suspensions more than 9 months old yielded protoplasts that failed to develop fertile plants. As a result, cultures less than 6 months old were used in plasmid uptake experiments. Aberations in chromosome number have been reported in root tips of transgenic rice, a significant number of the plants being triploid (Toriyama et al., 1988).

Inheritance

Mendelian inheritance predominates in all transgenic plants regenerated from electroporated protoplasts. Generally, primary transformants harboring an antibiotic-resistance gene give, upon self-fertilization, approximately 75% resistant progeny, which would be expected for a single dominant trait. In rape seed, 73% of the progeny seedlings were resistant (Guerche et al., 1987b), in rice 75% (Battraw and Hall 1992), and in tobacco 74–77% of progeny plantlets of most transformants were found to be resistant (Riggs and Bates, 1986, Table 10-2). Abberrant inheritance, however, also occurs. In kanamycin-resistant lettuce, the percentage of resistant progeny seedlings was as low as 52–55% for two selfed plants (Chupeau et al., 1989) whereas in tobacco, Riggs and Bates (1986) found three transformants where the resistant-to-sensitive ratio of the progeny was close to 2 : 1. Such a ratio is consistent with inheritance of a single lethal mutation in a homozygous locus, but other explanations are also possible.

High percentages (85–95%) of resistant progeny suggest independent inheritance of two or three functional copies of the introduced antibiotic resistance gene (Table 10-2; Shimamoto et al., 1989; Tada et al., 1990).

TRANSFORMATION OF INTACT CELLS AND TISSUES

Because of considerable difficulties associated with the regeneration of plants from protoplasts of some species, electroporation of intact cells and tissues have been performed, despite the general notion that the plant cell wall is impermeable to DNA. Data are now accumulating that show that some types of cells and tissues of some plant species do not exclude DNA.

Cells

Using mesophyll cells isolated by digesting tobacco leaves with 0.5% Macerozyme, Morikawa et al. (1986) were able to "electroinject" viral RNA into the cells.

Lindsey and Jones (1987b) obtained transient gene expression in intact sugar beet suspension cells, electroporated under conditions similar to those used for protoplasts. Enzymatic digestion with cellulases and pectolyases significantly increased transient expression, with 0.5% Pectolyase leading to the highest levels of expression. These data suggest that the pectin fraction rather than cellulose restricts DNA uptake through the cell wall.

Dekeyser et al. (1990) obtained transient expression in leaf bases of young etiolated rice plants that were washed extensively to remove nucleases (Sawicka, 1987). Plasmid DNA was incubated with the plant material 1 hr before electroporation. This method was also applicable to other monocotyledons, including wheat, maize, and barley.

Embryos

Dry viable embryos of some cereals and legumes have been shown to take up DNA from an imbibition solution and to express the DNA transiently. This phenomenon may be related to mechanical damage of the cell wall and special properties of membranes in dry embryos (Töpfer et al. 1989). Electroporation enhances DNA uptake significantly in embryos, particularly in maize embryos that do not take up DNA by imbibition. Electroporation of immature maize embryos resulted in transient expression as well as fertile transgenic plants. A short enzymatic treatment (1–3 min in 0.3% Macerozyme) or mechanical wounding was required. Embryogenic maize callus was also transformable by this procedure (D'Halluin et al., 1992). Compared to imbibition in DNA solution, electroporation of cowpea embryos considerably enhanced transient expression. This was further increased by adding compounds as spermine and lipofectin (Akella and Lurquin, 1993).

Microspores

Microspores can be permeabilized by electroporation to both low molecular weight dyes, such as propidium iodide, and to plasmid DNA (Joersbo et al., 1990, Abdul-Baki et al., 1990). Electroporated germinating microspores took up 6% of the plasmid in the medium while retaining 90% viability and their ability to fertilize emasculated flowers (Abdul-Baki et al., 1990). Transient expression has been obtained in electroporated maize (Fennell and Hauptmann, 1992) and tobacco pollen (Saunders et al., 1992).

SUMMARY

Among the methods available for gene transfer into plant cells, electroporation has proven to be an efficient and versatile technique. A variety of plant species, including economically important cereals, have been transformed by electroporation. This chapter presents a survey of the most significant parameters affecting gene delivery to plant cells by electroporation resulting in transient expression and stable transformation. Characteristics of transgenic plants produced by electroporation are also described.

References

Abdul-Baki, A. A., Saunders, J. A., Matthews, B. F., and Pittarelli, G. W. (1990). DNA uptake during electroporation of germinating pollen grains. Plant Sci. **70**: 181–190.

Akella, V., and Lurquin, P. F. (1993). Expression in cowpea seedlings of chimeric transgenes after electroporation into seed-derived embryos. Plant Cell Rep. **12**: 110–117.

Battraw, M., and Hall T. C. (1991). Stable transformation of Sorghum bicolor protoplasts with chimeric neomycin phosphotransferase II and β-glucuronidase genes. Theor. Appl. Genet. **82**: 161–168.

Battraw, M., and Hall, T. C. (1992). Expression of a chimeric neomycin phosphotransferase II gene in first and second generation transgenic rice plants. Plant Sci. **86**: 191–202.

Bekkaoui, F., Pilon, M., Laine, E., Raju, D. S. S., Crosby, L., and Dubstan, D. I. (1988). Transient gene expression in electroporated *Picea glauca* protoplasts. Plant Cell Rep. **7**: 481–484.

Bellini, C., Chupeau, M.-C., Guerche, P., Vastra, G., and Chupeau, Y. (1989). Transformation of *Lycopersicon peruvianum* and *Lycopersicon esculentum* mesophyll protoplasts by electroporation. Plant Sci. **65**: 63–75.

Benz, R., and Zimmermann, U. (1980). Relaxation studies on cell membranes and lipid bilayers in the high electric field range. Bioelectrochem. Bioenerg. **7**: 723–739.

Boston, R. S., Becwar, M. R., Ryan, R. D., Goldsbrough, P. B., Larkins, B. A., and Hodges, T. K. (1987). Expression from heterologous promotors in electroporated carrot protoplasts. Plant Physiol. **83**: 742–746.

Bower, R., and Birch, R. G. (1990). Competence for gene transfer by electroporation in a subpopulation of protoplasts from uniform carrot cell suspension cultures. Plant Cell Rep. **9**: 386–389.

Charest, P. J., Devantier, Y., Ward, C., Jones, C., Schaffer, U., and Klimaszewska, K. K. (1991). Transient expression of foreign genes in the gymnosperm hybrid larch following electroporation. Can. J. Bot. **69**: 1731–1736.

Christou, P., Murphy, J. E., and Swain, W. F. (1987). Stable transformation of soybean by electroporation and root formation from transformed callus. Proc. Natl. Acad. Sci. USA **84**: 3962–3966.

Chupeau, M.-C. Bellini, C., Guerche, P., Maisonneuve, B., Vastra, G., and Chupeau, Y. (1989). Transgenic plants of lettuce (*Lactuca sativa*) obtained through electroporation of protoplasts. Biol. Technol. **7**: 503–508.

Coster, H. G. L. (1968). The role of pH in the punch-through effect in the electrical characteristics of *Chara australis*. Aust. J. Biol. Sci. **22**: 365–374.

Coster, H. G. L., and Zimmermann, U. (1975). The mechanism of electrical breakdown in the membranes of *Valonia utricularis*. J. Membrane Biol. **22**: 73–90.

Dean, C., Jones, J., Favreau, M., Dunsmuir, P., and Bedbrook, J. (1988). Influence of flanking sequences on variability in expression levels of an introduced gene in transgenic tobacco plants. Nucl. Acids Res. **16**: 9267–9282.

Dekeyser, R. A., Claes, B., De Rycke, M. U., Habets, M. E., and Van Montagu, M. C. (1990). Transient gene expression in intact and organized rice tissues. The Plant Cell **2**: 591–602.

D'Halluin, K., Bonre, E., Bossut, M., De Beuckeleer, M., and Leemans, J. (1992). Transgenic maize plants by tissue electroporation. The Plant Cell **4**: 1495–1505.

Fennel, A., and Hauptmann, R. (1992). Electroporation and PEG delivery of DNA into maize microspores. Plant Cell Rep. **11**: 567–570.

Frearson, E. M., Power, J. B., and Cocking, E. C. (1973). The isolation, culture and regeneration of Petunia leaf protoplasts. Dev. Biol. **33**: 130–137.

Fromm, M., Taylor, L. P., and Walbot, V. (1985). Expression of genes transferred into monocot and dicot plant cells by electroporation. Proc. Natl. Acad. Sci. USA **82**: 5824–5828.

Fromm, M., Taylor, L. P., and Walbot, V. (1986). Stable transformation of maize after gene transfer by electroporation. Nature **319**: 791–793.

Guerche, P., Bellini, C., Le Moullec, J.-M., and Gaboche, M. (1987a). Use of a transient expression assay for the optimization of direct gene transfer into tobacco mesophyll protoplasts by electroporation. Biochimie **69**: 621–628.

Guerche, P., Charbonnier, M., Jouanin, L., Tourneur, C., Paszkowski, J., and Pelletier, G. (1987b). Direct gene transfer by electroporation in *Brassica napus*. Plant Sci. **52**: 111–116.

Hauptmann, R. M., Ozias-Akins, P., Vasil, V., Tabaeizadeh, Z., Rogers, S. G., Horsch, R. B., Vasil, I. K., and Fraley, R. T. (1987). Transient expression of electroporated DNA in monocotyledonous and dicotyledonous species. Plant Cell Rep. **6**: 265–270.

Hobbs, S. L. A., Jackson, J. A., Baliski, D. S., Delong, C. M. O., and Mahon, J. D. (1990). Genotype- and promotor-induced variability in transient β-glucuronidase expression in pea protoplasts. Plant Cell Rep. **9**: 17–20.

Horn, M. E., Shillito, R. D., Conger, B. V., and Harms, C. T. (1988). Transgenic plants of orchard grass (*Dactylis glomerata* L.) from protoplasts. Plant Cell Rep. **7**: 469–472.

Huang, Y.-W. and Dennis, E. S. (1989). Factors affecting stable transformation of maize protoplasts by electroporation. Plant Cell, Tissue and Organ Cult. **18**: 281–296.

Joersbo, M. (1990). Methods for direct gene transfer into plant protoplasts. Ph.D. Thesis, University of Aarhus, Denmark.

Joersbo, M., and Brunstedt, J. (1990a). Direct gene transfer to plant protoplasts by electroporation by alternating, rectangular and exponentially decaying pulses. Plant Cell Rep. **8**: 701–705.

Joersbo, M., and Brunstedt, J. (1990b). Quantitative relationship between parameters of electroporation. J. Plant Physiol. **137**: 169–174.

Joersbo, M., Jorgensen, R. B., and Olesen, P. (1990). Transient electropermeabilization of barley microspores to propidium iodide. Plant Cell Tissue Organ Cult. **23**: 125–129.

Joersbo, M., and Brunstedt, J. (1991). Electroporation: Mechanism and transient expression, stable transformation and biological effects in plant protoplasts. Physiol. Plant. **81**: 256–264.

Jones, H., Ooms, G., and Jones, M. G. K. 1989. Transient gene expression in electroporated Solanum protoplasts. Plant Mol. Biol. **13**: 503–511.

Kirches, E., Frey, N., and Schnabl, H. (1991). Transient gene expression in sunflower mesophyll protoplasts. Bot. Acta **104**: 212–216.

Köehler, F., Golz, C., Eapen, S., Kohn, H., and Schieder, O. (1987a). Stable transformation of moth bean *Vigna aconitifolia* via direct gene transfer. Plant Cell Rep. **6**: 313–316.

Köehler, F., Golz, C., Eapen, S., and Schieder, O. (1987b). Influence of plant cultivar and plasmid-DNA on transformation rates of tobacco and moth bean. Plant Sci. **53**: 87–91.

Krens, F. A., Molendijk, L., Wullems, G. J., and Schilperoort, R. A. (1982). In vitro transformation of plant protoplasts with Ti-plasmid DNA. Nature **296**: 72–74.

Lindsey, K., and Jones M. G. K. (1987a). The permeability of electroporated cells and protoplasts of sugar beet. Planta **172**: 346–355.

Lindsey, K., and Jones, M. G. K. (1987b). Transient gene expression in electroporated protoplasts and intact cells of sugar beet. Plant Mol. Biol. **10**: 43–52.

Lindsey, K., and Jones, M. G. K. (1989). Stable transformation of sugar beet protoplasts by electroporation. Plant Cell Rep. **8**: 71–74.

Lurquin, P. F., and Paszty, C. (1988). Electroporation of tobacco protoplasts with homologous and non-homologous transformation vectors. J. Plant Physiol. **133**: 332–335.

Morikawa, H., Iida, A., Matsui, C., Ikegami, M., and Yamada, Y. (1986). Gene transfer into intact plant cells by electroinjection through cell walls and membranes. Gene **41**: 121–124.

Oard, J. H., Paige, D., and Dvorak, J. (1989). Chimeric gene expression using maize intron in cultured cells of breadwheat. Plant Cell Rep. **8**: 156–160.

Ou-Lee, T.-M. Turgeon, R., and Wu, R. (1986). Expression of a foreign gene linked to either a plant-virus or a *Drosophila* promotor, after electrooration of protoplasts of rice, wheat and sorghum. Proc. Natl. Acad. Sci. USA **83**: 6815–6819.

Percival, F. W., Cass, L. G., Bozak, K. R., and Christoffersen, R. E. (1991). Avacado fruit protoplasts, a cellular model system for ripening studies. Plant Cell Rep. **10**: 512–516.

Potter, H. (1988). Electroporation in biology: Methods, applications and instrumentation. Anal. Biochem. **174**: 361–373.

Puonti-Kaerlas, J., Ottosson, A., and Eriksson, T. (1992). Survival and growth of pea protoplasts after transformation by electroporation. Plant Cell, Tissue and Organ Cult. **30**: 141–148.

Rathus, C., and Birch, R. G. (1992a). Optimization of conditions for electroporation and transient expression of foreign genes in sugarcane protoplasts. Plant Sci. **81**: 65–74.

Rathus, C., and Birch, R. G. (1992b). Stable transformation of callus from electroporated sugarcane protoplasts. Plant Sci. **82**: 81–89.

Rhodes, C. A., Pierce, D. A., Mettler, I. J. Mascarenhas, D., and Detmer, J. J. (1988). Genetically transformed maize plants from protoplasts. Science **240**: 204–207.

Riggs, C. D., and Bates, G. W. (1986). Stable transformation of tobacco by electroporation. Proc. Natl. Acad. Sci. USA **83**: 5602–5606.

Rouan, D., Montané, M.-H., Alibert, G., and Teissié. (1991). Relationship between protoplast size and critical field strength in protoplast electropulsing and application to reliable DNA uptake in Brassica. Plant Cell Rep. **10**: 139–143.

Saunders, J. A., Matthews, G. F., and Van Wert, S. L. (1992). Pollen electrotrans-formation for gene transfer in plants Pages 227–247. In Guide to Electroporation and Electrofusion. Chang, D. C., Chassy, B. M., Saunders, J. A., and Sowers, A. E. eds. Academic Press, New York.

Sawicka, T. (1987). Membrane-bound nucleotic activity of corn root cells. Phytochem. **26**: 59–63.

Seguin, A., and Lalonde, M. (1988). Gene transfer by electroporation in betulaceae protoplasts: *Alnus incana*. Plant Cell Rep. **7**: 367–370.

Shillito, R. D., Saul, M. W., Paszkowski, J., Mueller, M., and Potrykus, I. (1985). High efficiency direct gene transfer to plants. Biotechnol. **3**: 1099–1103.

Shimamoto, K., Terada, R., Isawa, T., and Fujimoto, H. (1989). Fertile transgenic rice plants regenerated from transformed protoplasts. Nature **338**: 274–276.

Tada, Y., Sakamoto, M., and Fujimura, T. (1990). Efficient gene introduction into rice by electroporation and analysis of transgenic plants: Use of electroporation lacking chloride ions. Theor. Appl. Genet. **80**: 475–480.

Tagu, D., Bergounioux, C., Perennes, C. and Gadal, P. (1990). Inheritance of two foreign genes co-introduced into *Petunia hybrida* by direct gene transfer. Plant Cell, Tissue and Organ Cult. **21**: 259–266.

Tautorus, T. E., Bekkaoui, F., Pilon, M., Datla, R. S. S., Crosby, W. L., Fowke, L. C., and Dunstan, D. I. (1989). Factors affecting transient expression in electroporated black spruce (*Picea mariana*) and jack pine (*Pinus banksiana*) protoplasts. Theor. Appl. Genet. **78**: 531–536.

Töpfer, R., Gronenborn, B., Schell, J., and Steinbiss, H.-H. (1989). Uptake and transient expression of chimeric genes in seed-derived embryos. The Plant Cell. **1**: 133–139.

Toriyama, K., Arimoto, Y., Uchimiya, H., and Hinata, K. (1988). Transgenic rice plants after direct gene transfer into protoplasts. Bio/Technol. **6**: 1072–1074.

Tsukada, M., Kusano, T., Kitagawa, Y. (1989). Introduction of foreign genes into tomato protoplasts by electroporation. Plant Cell Physiol. **30**: 599–603.

Tyagi, S., Spoerlein, B., Tyagi, A. K., Herrmann, R. G., and Koop, H. U. (1989). PEG- and electroporation-induced transformation in *Nicotiana tabacum*: Influence of genotype on transformation frequencies. Theor. Appl. Genet. **78**: 287–292.

Widholm, J. M., Dhir, S. K., and Dhir, S. (1992). Production of transformed soybean plants by electroporation of protoplasts. Physiol. Plant. **85**: 357–361.

Wilson, C. M. (1968a). Plant nucleases. I. Separation and purification of two ribonucleases and one nuclease from corn. Plant Physiol. **43**: 1332–1338.

Wilson, C. M. (1968b). Plant nucleases. II. Properties of corn ribonucleases I and II and corn nuclease I. Plant Physiol. **43**: 1339–1346.

Wirtz, U., Schell, J. and Czernilofsky, A. P. (1987). Recombination of selectable marker DNA in *Nicotiana tabacum*. DNA **6**: 245–253.

Zhang, H. M., Yang, H., Rech, E. L., Golds, T. J., Davis, A. S., Mulligan, B. J., Cocking, E. C., and Davey, M. R. (1988). Transgenic rice plants produced by electroporation-mediated plasmid uptake into protoplasts. Plant Cell Rep. **7**: 379–384.

Zimmermann, U., and Benz, R. (1980). Dependence of the electrical breakdown on the charging time in *Valonia utricularis*. Membrane Biol. **53**: 33–43.

Zimmermann, U., Pilwat, G., and Riemann, F. (1974). Dielectric breakdown of cell membranes. Biophys. J. **14**: 881–899.

CHAPTER 11

Electromagnetic Cell Stimulation

D. Jones

B. McLeod

ABSTRACT

Electromagnetic field (ELF) stimulation of biological tissues has been, and remains, a contentious issue in the scientific world. Many of the experiments have not been reproducible in other laboratories, and many appear not to have been thoroughly conducted. Several experiments, however, do appear to meet these criteria, and the data from these experiments should be taken seriously. The physical laws behind these ELF effects are unclear. Until a theoretical physical framework is constructed, the use of electromagnetic fields experimentally will be fraught with methodological dangers. It seems that new ideas in physics and new approaches in biology are needed to account for most of the biological effects reported. Thus, the construction of artifact-free experimental systems will be difficult. A critical review of the current theories of action is presented.

INTRODUCTION

Do weak magnetic fields, low-energy electrical fields, electromagnetic radiation (low-frequency radio waves or ELF), amplitude modulated microwaves, AC and DC electrical field, frequency- and amplitude-modulated ELF, and time varying magnetic fields (PEMF or pulsed electro-magnetic fields) affect biological systems? These are questions raised over the last century by clinicians and scientists from many backgrounds.

At the present time, the general scientific community has not identified a mechanism that can explain how low-energy magnetic fields can have biological effects (what is considered here as a low energy is less than 0.5 eV incident energy per cell, about 1 eV nm^2, for time-varying fields and a static magnetic field strength of less than 10^{-2} T). Many do not accept the possibility of such effects.

Much of the literature describing these effects is to be seen in the transactions of societies founded to provide a forum for the publication of such reports. Access to the literature through abstracting journals is not often possible. Also, many of the reports in the literature are not repro-ducible. One example of this was a study on nor-adrenaline release in a neural cell line (Dixey and Rein, 1982), as admitted by one of the authors several years ago. This problem of reproducibility has also affected studies on second messenger systems (cAMP, protein kinases, and free intracellu-lar calcium) (Jones, 1984; Jones and Ryaby, 1987; Jones et al., 1986b), which are variable enough when studying "normal" biology.

Put in another way, the problem is one of how incident energies of much less than the thermal energy kT stimulate a process unless there is some receptor, an amplification, or time integration mechanism. For instance, in the case of vibration sensors in insects, the sensory cells have been shown to be sensitive to a sound energy input of less than 0.3 eV per cell due to a structure designed to receive and amplify such energies (Thurm, 1983). Therefore the question is What are the receptors?

It is important to be critical at the present time, since some groups claim that ELF such as found near radio-transmitting stations or fields found in the region of high-tension power lines (50 or 60 Hz, depending on the country) can cause health problems, an issue that is hotly discussed (Foster and Pickard, 1987). The U.S. Environmental Protection agency has only recently opened its mind to the possibility of such an effect, although the U.S. Federal Drug Administration has approved ELF type (PEMF) devices for use in certain therapies (Bassett et al., 1981).

General Considerations

If we suppose a physical interaction has a significant general effect, for instance, it can in some way affect a dipole moment of a protein, then we cannot expect to see a specific biological response. For instance, warming an organism from its lower physiological temperature range to its higher temperature range usually has the effect of increasing all enzymatic activity and increasing metabolism.

A second point is that the site of interaction must be observed in real time, that is, during the exposure. Responses measured after a week of exposure may or may not be due to the stimulation applied. Responses to fields that require days of integration to have an effect must be shown to be the result of integration of energy through some mechanism and not due to slight modification.

Third, in frequency-dependant effects, the interaction must reflect some property of the biological system in terms of the time of interaction. Time-varying magnetic fields and modulated microwave fields are said to be effective at frequencies of about 15 Hz, which implies an interaction with a process having a time scale of 66.6 msec. If the biological targets have relaxation times much shorter than this (such as ions moving through a channel at 1 μsec or ions in a pump at 1 msec), then the relaxation times do not allow time for interaction with the field. This observation can also be extended to static magnetic and electrical field effects—it will be the long-term (more than 1 sec) biological processes that will be most affected if there is some interaction.

The reported affects appear to result in specific responses of the cells and organisms and do not appear to stimulate all activities or even some activities all of the time. Any hypothetical physical mechanism(s) must explain the specific responses reported.

Biological Transduction

Many processes have been reported to be affected in many different organisms. How are these results to be evaluated? Do the effects on differentiation [e.g., in melanoma cells, (Jones et al., 1986b) in osteoblasts (Murray and Farndale, 1985), and in insects (Goodman et al., 1983)] measured at different biochemical levels and time after exposure form part of the same transduction mechanism as the expression of heat shock-like proteins in insect salivary glands?

One possible approach to analyze the site of interaction is to look for the first significant biochemical response. The earlier something happens after exposure, the closer it must be to the transduction mechanism. It will not do to investigate calcium dynamics after a period of 30 min, since many other processes could in the meantime be activated and responsible for the change. A further difficulty is that many biological mechanisms are highly interactive and feedback both positively and negatively on one another.

Analyses made after several hours of exposure might be the result of many interacting processes. Measurement of the cAMP dependant protein kinase (PKA) activity after 2 min exposure to PEMF by Jones and Ryaby (1987) is a relatively rapid measurement as opposed to measurements lasting hours, days, and weeks. These changes in activity are more likely to be closer to the original site of interaction than processes measured even after several minutes. As an example, measurements of osteoblast membrane potential during PEMF exposure showed that although the membrane potential hyperpolarized after 5 min, this could easily be due to another mechanism.

The changes in PKA activity were considered to control the response of the cell line studied, since a specific inhibitor of PKA blocked PEMF response but not insulin kinase-induced effects. Byus et al. (1984) reported on the change in activity of an uncharacterized kinase after 15 min of exposure to modulated microwave fields, but found no change in PKA activity. In the PKA study above, PKA activity had returned back to nearly control levels by 15 min, as is in the case of hormonal stimulation. Many biological "trigger" mechanisms last only a few seconds, and not measuring at the right time could miss the effect.

These kinase changes themselves might not be the target for interaction, since their activation are themselves the product of earlier activation. The receptor itself is not considered to be the target, since destroying the receptor with N-ethylmaleimide does not affect the PEMF response but destroys the hormone response. The only possible targets for PEMF effect left are the enzymes responsible for transducing the hormone effect into kinase activation, the G regulatory proteins (Jones et al., 1987), or the calcium ion. Of these, the G regulatory proteins and the calcium ion, for reasons discussed below, are therefore the likely targets. It is therefore suggested that many of the effects of PEMF reported can be due to either altered G protein activity or a change in calcium ion concentration.

Methodological Problems

The exposure of biological experiments to fields is fraught with problems. Normal incubating systems are made of metal, which for many experiments is not suitable. The most convenient plastic to use is Perspex (acrylic sheet). This type of plastic, however contains toxic monomers. Polycarbonate usually doesn't suffer from this problem. Current used in coils can, in certain circumstances, cause heating, and in any coil system, vibration can be a problem due to magnetic repulsion. Vibration can reduce cells settling on the substrate and alter gas exchange in the medium. Standing waves in the medium as a result of vibration have also been reported to cause cells to grow at the nodal point.

Several workers use biological systems as if they are easy to set up. In a number of reported experiments, the non-exposed control does not appear to give the normal expected values. This is especially so in the reports of malformation of chick embryos exposed to PEMF. In all of these cases, the control incubator had an extremely high percentage of malformed embryos, more than 19%. In a well set up incubator a value of less than 1% is to be expected. Rooze and Hinsenkamp (1985), investigating quail embryo development, only noticed an accelerated development in the early stages, but hatching was at the normal time and the chicks quite normal. In Münster we have reproduced this work using Leghorn fowl. Using ELF fields in a commercial incubator, the embryos at 14 days were 2 stages accelerated in development and weighed 33% more than the controls. Hatching, however, took place at 21 days as in the controls, and both groups were the same. In these experiments, although a "biological" effect was observed, no malformation of the chicks was noticed. In the case of the commercial incubator studies, the "no hatch" rate was 1.7% in both groups, due to infertile eggs and 0% malformation in 760 embryos studied.

It therefore seems that, at best, results can be obtained when the biological material is under certain conditions, although what these are is not known at present.

Variability

Although variability and reproducibility in cell biological data is often not as good as can be wished for, and often not as good as is published, the variability in PEMF experiments is often much higher. As an example of this, the effect of PEMF on 864 chick limb organ culture tibial cartilage DNA content and 468 control limb cartilages is reported. The DNA content in the control was 83.6 ± 7 μg and in the PEMF treated, 82.7 ± 13 μg. This is not significantly different by non-parametric tests of significance. An F test on the distribution of the control group showed no significant difference from a normal distribution. The PEMF-treated group, however, showed a highly significant deviation from the normal distribution ($p < 0.000001$). Analysis of the results showed the presence of three separate groups, each normally distributed, 40% with an average DNA content of 74.6 μg, a small group with an average DNA content of 83.2 μg, and a third group of 38% with an average content of 87.5 μg. Although the overall averages were not different, the change in distribution due to significantly higher variation seems to suggest that PEMF had no effect on 20%, reduced the growth of 40%, and stimulated the growth of the other 40% of tibiae.

As mentioned above, variation can also be caused by the experimental conditions, which does not make the task of evaluating the experiments easier.

Static Magnetic Field Effects

Interestingly, supporters of PEMF, AC fields, and ELF work seem to disregard the possibility of direct magnetic field effects, although there are many such reports in the specialized literature for other effects.

In some cases the "receptor" for the magnetic field is clear. Blakemore (1975), for instance, described a structure resembling a ferro-magnet in a specialized organelle of certain bacteria. Similar organelles have been found in a variety of other animals and are part of the navigation system of dolphins and pigeons. Baker et al. (1983) have reported similar organelles in humans. Beischer (1968) and Conley (1969) reviewed static and pulsed magnetic field effects nearly 30 years ago.

Conley (1969) has given a short review on this and also discussed methods to counter the Earth's magnetic field variability. Some experiments have been conducted using suitable precautions, and many others have not. One of the most intensely investigated phenomena in magnetic field interaction is the effect on the eyes of invertebrates and vertebrates. Arendse (1978) reported the detection of magnetic fields by the light detection organ in worms. Raybourn (1983) showed DC magnetic field responses in turtle retinas. Reuss and Olcese (1986) and Reuss et al. (1984) reported magnetic field effects in rat pineal glands and cells. These are a selection of thousands of papers written for specialist journals.

In investigating the effect of PEMF on bone cells, Jones and Ryaby (1987) used permanent magnets as a control for studying the effects of time-varying magnetic fields. It was a surprising result that the static magnetic fields (between 10^{-2} and 10^{-3} T) caused the same response as PEMF and the results were less variable. Bruce et al. (1987) reported that similar field strengths increased the fracture healing strength of callus in rabbits. The mechanism of interaction of weak static magnetic fields has been much discussed. Weller et al. (1984) considered effects on geminate radical production interactions at nanosecond times of reaction in fields between 0.3 and 2×10^{-2} T. The radicals were produced by light, and thus this theory is most applicable to the effect of magnetic fields on the light reception organs above. This work demonstrates that although the strength of the magnetic interaction is much smaller than the thermal energy kT, magnetic fields can perturb a coherent quantum mechanical process. These results can be reproduced (Shulten and Wolynes, 1978; Shulten, 1982).

Shulten et al. (1986) used these results to consider biological implications of the reactions studied and showed chemical reaction modifications at 10^{-4} T. Long time reactions such as those occurring in the GTP-binding regulatory proteins were considered to be a possible site of interaction—a result that could explain the sensitivity of the light detection organs (in vertebrates eyes, but not in invertebrates) as mentioned above. The effects described by these authors are related in mechanism to nuclear magnetic resonance (NMR) and to the Zeeman effect (quantum splitting). No test of this hypothesis in a complete biological system has, however, been undertaken to see which component is sensitive to these reactions. It should be noted that usually experiments are not performed in a homogenous or static magnetic environment due to variations in the Earth's magnetic field. Many authors, for example, Blackman et al. (1985a, b), suggest that the Earth's magnetic field plays a role in interaction of fields with the biological tissue. In view of this, experiments not controlling for this factor might not be valid or might play a role in causing the higher variability noticed in many of the experiments.

DC and AC Electrical Fields

The biological effects of low-energy DC and AC fields are discussed by Goldsworthy in chapter 12 of this book. A brief discussion in relation to the EMF question is in order here, however, since one possible mechanism of action is due to the induction of electrical current by time-varying magnetic fields.

It is known that members of the shark family are sensitive to weak electrical fields in the order of 0.2 μV/cm and 2 μA at low frequency (5 Hz) or DC currents. The isolated ampullae themselves show a similar sensitivity. Cells can be demonstrated to exhibit galvanotaxis in very low currents and voltages, from between 0.1 mV/cm (Sisken and Smith, 1975) and 10 mV/cm (Jaffe and Poo, 1979). Since certain isolated cells seem to be very sensitive, theories that presuppose gap junction circuits to form antennae (Cooper, 1984) do not seem called for. Cells connected by gap junctions are usually connected electrophysiologically at about 20% conduction.

It is clear that weak electrical fields do have demonstratable biological activity in certain cases. In the case of time-varying magnetic fields, discussed below, the induction of weak currents in the tissue might therefore play a role in any of the responses described. A method to distinguish experimentally between the induced electrical current and the magnetic field is also described below.

Amplitude Modulated Radiofrequency Electromagnetic Radiation

The biological effects of high-frequency (above 100 MHz) amplitude modulated at low frequency (16 Hz) were first investigated by Bawin et al. (1978) on calcium afflux from chick brain tissue in 1978 (Bawin et al., 1975; Bawin and Adey, 1976) measured after 10 min. The incident energy used was about 0.75 W/cm^2, and a calculated voltage of 1 $\mu V/cm$ was induced in the tissue. It may be that the biological effects are due to the same processes as discussed below for time-varying magnetic fields.

Pulsed Magnetic Fields

PEMF was introduced in the United States by Bassett et al. (1974) as a way to reintroduce strain related potentials (SRP) into bone. SRPs were first discovered by Fukada and Yasuda (1957), and at first it was thought that the origin was piezo electrical, but more recently the contribution of streaming or zeta (ζ) potentials (Digby, 1966) has been realized. Bassett and Becker (1962) suggested that the SRPs were physiologically active and indeed were responsible for bone remodeling. Although more than 200,000 patients have been treated with this system to date, some clinicians, especially in Europe, do not accept that PEMF can help heal non-united bones (British Medical Journal Secretariat, 1980; Mulier and Spaas, 1980; Barker et al., 1984). A double blind trial for rotator cuff tendonitis (Binder et al., 1984) is the most acceptable clinical trail for showing a highly positive effect. Animal and cell culture experiments investigating the effect of PEMF on bone metabolism do not show agreement as to whether such a positive effect exists. For instance, Law et al. (1985) investigating sheep, Smith and Nagel (1983) in rabbits, Yamada et al. (1985) in cultured cells, all failed to find evidence of any effect. On the other hand, Bassett using a variety of animal models (Bassett et al., 1979a, b; Bassett and Hess, 1984), together with Farndale and Murray (1985); Murray and Farndale (1985); Luben et al. (1982), and Fitzsimmons et al. (1986), have all reported either significant or slightly significant effects. A skeptic would think that significant effects are found most by groups with an interest in finding one, and no significant results are reported from groups who are negative to the idea of there being an effect.

From the literature it seems that in nearly every conceivable organism from bacteria to humans an isolated organ has been exposed to PEMF to look for biological effects. In a large number of cases the reports are conflicting, but, as discussed below, variations in frequency, energy, and dosage (amount of exposure and pattern of exposure) can cause this. One problem with using a common PEMF generator, the Electro-Biology Incorporated (EBI) generator, is that the frequencies applied are numerous and the method of "chopping" the wave-forms is not exact so that a wave-form is often started halfway through a cycle. This means that the exact frequencies are difficult to measure and vary throughout the exposure.

Assuming that some of the reports are not "noise" and that real biological effects are being observed, what is the mechanism of interaction? Given that biological effects do occur, one question that requires answering is does ELF work through the magnetic field or through the induced electrical field. That weak magnetic fields have effects (at least on chemical reactions) appears fairly sound. Electrical fields have demonstrable effects, the least contentious of which is the demonstration of galvanotaxis. One way to distinguish the electrical field effect from the magnetic field effect using ELF in cell culture is to construct a Petri plate consisting of concentric rings isolated from one another. When the plate is put between the two Helmholtz coils, the electrical field will be strongest in the outside ring and weakest in the middle one, but the magnetic field strength will be the same over the whole plate. Unpublished experiments conducted with such a plate indicate a role for the induced electrical field in producing a biological response. This point is important when considering possible mechanisms of action since many current hypotheses are concerned with the action of the time-varying magnetic field (see below).

Several types of hypotheses have been discussed to overcome the kT energy problem above and involve an antenna effect or an amplifying effect or depend on a resonator to bring about chemical reactions. Reception of radio and television programs are familiar to everyone, even though the energy density is much lower than that under discussion here. Radio waves are received because of the antenna principle where a structure intercepts the energy over some surface. The receiver contains a resonant circuit tuned to the proper frequency and the resonant current is then amplified. This principle of amplification and resonance is used by many workers in the form of many hypotheses, such as the cyclotron resonance hypothesis (Liboff, 1985), where the receptor is an ion, by the gap junction hypothesis (Cooper, 1984), where the length of a biological antenna is increased by supposed electrical connections, and where a protein acts as a receptor, either as an antenna (Adey, 1981) or as a resonator (Chiabrera and Bianco, 1987). These hypotheses are discussed further below.

POSSIBLE MECHANISMS OF ACTION OF ELF AND PEMF

An often cited result obtained by Bawin et al. (1975, 1976, 1978a, 1978b) showed that calcium efflux from chick cerebral tissue could be significantly influenced by weak, sinusoidally oscillating electromagnetic fields applied with a high-frequency carrier wave. The work suggested a resonance or frequency window since the efflux not only varied as the frequency of the ELF signal was varied, but also there was a distinct maximum efflux at a specific frequency. This work was later repeated by Blackman et al. (1979, 1982, 1985a, 1985b), who obtained an interesting piece of additional data. In their experiments, the calcium efflux was increased or decreased as a function of the frequency of the applied ELF field, and also the efflux was varied as a function of the magnitude of the DC (non-time-varying) magnetic field to which the tissue was exposed. The DC magnetic field vector was perpendicular to the AC field vector for this work. A series of papers was published (McLeod et al., 1987; Smith et al., 1987; Liboff et al., 1987, 1990) that suggested an ion-specific resonance effect when the ELF AC field and the DC magnetic field vectors were parallel and combined according to the expression

$$f_c = \frac{1}{2\Pi}\left(\frac{q}{m}\right)(B_0) \tag{1}$$

where

f_c = the applied AC frequency
(q/m) = the ion charge to mass ratio (coulombs per kilogram)
B_0 = the DC magnetic field (tesla)

This work also suggested a distinct set of harmonics (integer multiples of f_c) that could affect the biological system (Liboff et al., 1990). Certain specific harmonics were also effective in the work by Blackman et al. (1985a, 1985b).

Goodman et al. (1983) observed cellular transcription induced by pulsed electromagnetic (EM) fields, and this work was followed by a series of papers (Goodman and Henderson, 1986a, 1986b, 1987a, 1987b) that reported transcription alteration in cells using both pulsed and sinusoidal PEMF fields. No DC fields were present except the ambient geomagnetic field. This work has been successfully reproduced in detail in other laboratories.

In addition to accounting for the above-reported effects, physical theories or models must also address the fact that the energy input of most of the PEMF and ELF fields used to observe biological effects is well below the background thermal energy (kT) level of the biological system. This point has been discussed by Adey (1988) since 1975 and was recently considered at length by Weaver and Astumian (1990). The Weaver and Astumian work suggests that any mechanism that is proposed to explain PEMF and ELF biological system interaction must contain the provision for either a narrow band response (a biosystem response within a small range of frequencies), or the theory must provide for a method of signal averaging (i.e., signal integration over a period of time) to avoid being found invalid based on the thermal noise limit. These two possibilities may also provide one key to testing physical theories; that is, by measuring the minimum time it takes for a biological system to respond to an PEMF or ELF signal, it should be possible to tell if narrow banding or signal averaging is taking place.

It has been accepted for quite some time that PEMF and ELF fields can elicit growth response in bone and bone cells (Bassett et al., 1982; Luben et al., 1982). These early papers also contained an additional piece of data. Adey (1981) suggested that the state of the biological system was also important. That is, he suggested that if the cell system was undergoing any kind of dynamic biological process, it was more probable that the biological system would respond to an applied PEMF or ELF field. A biosystem exactly in homeostasis would probably have sufficient feedback mechanisms active in order to prevent an observable response to a low-energy PEMF or ELF signal. The early non-union bone papers (Bassett et al., 1982) contained these data. Only the injured bone, not the normal bone (and tissue) that was also exposed to the PEMF, responded to the PEMF signal. Therefore, an acceptable physical theory must also address this important point.

The following pages review several of the theories or models that have been put forth. Some of the models address most of the above points; others address only a few of the points. This could suggest more than one mechanism must be considered in the ultimate model, or some of the models may address only a portion of the entire mechanism. The extent to which any of the theories can address the data may suggest a means of testing some of the proposed models.

There seems to be, for some unknown reason, a consensus that PEMF and ELF fields interact with biological systems at the level of the membrane surrounding cells in the biosystem. Most of the models and theories that have been proposed in recent years postulate some action that involves the signal and the lipid bilayer via receptors on or near that layer, intramembranous particles (IMPs), channels through the membrane, binding sites inside the membrane, or a combination of several of the above.

Usually these models have been proposed, then modified over some period of time as more experimental evidence has been accumulated. This chapter discusses the latest version of a particular theory.

Receptor Antenna Hypothesis

The model proposed by Adey (1990a, 1990b) is perhaps as complete as any of the several currently being considered by the research community. This model suggests that the ELF energy is detected at the surface of cells in tissue by strands of protein that extend above, through, and inside the cell membrane. Due to the energetic states of populations of electrical charges on these strands of proteins, an amplification of the weak ELF signal can occur if the populations of charges act in a cooperative fashion. In effect, the model suggests that the ELF energy can set in motion some action within the cell by triggering a cascade of action outside the cell but within the populations of charges on the protein strands. These strands then transmit the cascade action into the cell interior via the transmembrane portion of the protein strands. Once the "signal" has been transmitted to the cell interior, it is presumed to couple to vital cell processes to alter some action of the cell. It is repeatedly pointed out in discussions of the model that the biological system must be in a state far from equilibrium (Adey, 1990b) for the weak ELF signal to be detected and amplified into an observable result. It should be noted that if the cascade or amplification does occur, this implies that the initial, weak signal eventually becomes a much more energetic signal through the process of cooperative energy release at many binding sites. For this model, then, the cascade provides a means to answer the question of how the weak ELF signal could be detected by the biological system even if the ELF field energy level is below the thermal (kT) energy level. The general outline of the model as stated above is relatively straightforward. The details of the model are, however, not completely clear.

Soliton Hypothesis

It is postulated that once the PEMF or ELF signal is detected and the amplification process or cascade has begun, a signal is somehow transmitted into the cytoplasm. One suggested method of transmittal in this model is that ion flow through the intermembranous region would be enhanced by soliton waves. Although soliton waves are relatively well understood, an effort to find these waves in helical proteins and DNA has not proven their existence in biological systems (Lawrence et al., 1987).

Membrane Double Layer

A model has been proposed by Blank (1987) that has as its focus the cell membrane surface and the electrical double layer at that surface. This model makes use of the Surface Compartment Model (SCM) (Blank, 1982, 1983, 1984, 1987; Blank and Kavanaugh, 1982; Blank et al., 1982) in an attempt to provide a description of ionic movement through the membrane as a function of the applied ELF field. When the model is used to analyze an SCM voltage-controlled channel to which an AC field has been applied, it is found that surface ionic concentrations can reach new steady-state values that are functions of the AC frequency of the applied field. These apparent changes in the electrical double layer do not average out as might be expected with an applied field that is alternately positive then negative, but instead they build up over time. This implies an integration effect where small electrical fields might have a large effect after some period of time. It is not apparent just how long the time must be, but this model attempts to overcome the kT limitation by using this additive effect. The model does not address the question of how a DC magnetic field could modify the biological effects predicted by the modulation of the double layer. It is also not clear how this model would explain why only biological systems that are not in equilibrium would tend to respond to the applied AC electrical field. It should be noted that this theory is based upon the electrical field interacting with the double layer, and that it is an "integration" type of model.

Lorentz Force and "Cyclotron Resonance"

Several different models have been proposed that attempt to use the Lorentz force equation to calculate the effect of magnetic fields on charged particles near or in the cell membrane. When a charged particle such as an ion moves in a DC magnetic field, a force is exerted on the ion that tends to change the direction of travel of the ion. Most of the theories involving the Lorentz force equation have resulted from attempts to explain the experimental results obtained by Blackman (1985b), Liboff et al. (1987, 1990); McLeod et al. (1987); Smith et al. (1987) and Chiabrera and Bianco (1987). All of these experiments indicated that the biological effect observed when cells or tissues were exposed to ELF magnetic fields could be changed by changing the DC magnetic field at the exposure site. The latter experiments (Chiabrera and Bianco, 1987; Liboff et al., 1987, 1990; McLeod et al., 1987; Smith et al., 1987) were expressly designed to test for an ion-specific resonant effect of the DC magnetic field. The several theories discussed below can be grouped into a general category called cyclotron resonance (CR) models since the expression of Equation 1 eventually appears in all the theoretical work, and the equation is known as the cyclotron resonance expression. The reason Equation 1 emerges is a consequence of invoking the Lorentz force on the moving charged particle.

The use of the words "cyclotron resonance" to describe these models has created considerable controversy (Bioelectromagnetics Society Newsletter, 1990) due, in part, to the fact that cyclotron resonance can be given a specific physical meaning when one is describing a charged particle moving in a DC magnetic field in a vacuum. This, obviously, is not the case for a charged body moving in a biological system. Again, it should be pointed out that Equation 1 appears in the mathematics as a result of the starting point (Lorentz force) and since it is a convenient notation to work with in any development.

Note again that experiments have been done that have been specifically designed to look for resonance effects using the conditions required by Equation 1 (Carson et al., 1990; Chiabrera and Bianco, 1987; Liboff et al., 1987, 1990; Lyle et al., 1989; McLeod et al., 1987; Ross, 1988; Smith et al., 1987; Walheczek and Liburdy, 1990; Bioelectromagnetics Society Newsletter 1990). These data indicate that ion-specific resonance is occurring at the CR frequency, parallel AC and DC magnetic fields are required, and no biological effect occurs with just the AC or just the DC field applied. Thus, no matter how the results are named, it appears that AC ELF magnetic fields can be coupled to biological systems by using a specific DC magnetic field applied simultaneously in a specific way. The next few pages briefly present some current CR theories.

Chiabrera and Bianco (1987) have focused on the motion of a charged ligand at its binding site. In this work, the ligand is defined to be any charged body (ion, lectin, hormone, or antigen) that may be trapped for some period of time at a binding site. It is suggested that the ELF field acts to modulate the interaction of the ligand with the binding site that results in the production of a second messenger. The second messenger then alters some biological process. In essence, the model describes the action as ELF coupling to a ligand, subsequent modification of the complex velocity of the ligand within the microenvironment of the binding site, the production of a second messenger, and finally a biological modification. In this theory, as in most of the others that will be discussed, a second physical barrier must be addressed [the first being the thermal (kT) limit]. In virtually all of the experiments referenced above, a biological response was obtained when the frequency of the AC magnetic field was set by using the ion charge to mass ratio as shown in Equation 1. The mass was taken to be the mass of the ion with no solvation shell (e.g., a "bare" ion). Critics of the CR theories immediately pointed out that charged ions in solution always had to have a water shell, which would greatly increase the effective ionic mass and thus lower the resonant frequency substantially. The experimental data are not in agreement with this point. It could be considered that in the microenvironment of a binding site, the charged ligand can lose its solvation shell and could also experience a relatively collision-free environment. After making this point, Chiabrera and Bianco (1987) proceed to model the reaction between the ligand and its binding site as being modified by the magnetic component of the Lorentz force. To obtain a result that can be examined analytically (eventually numerically), several simplifying assumptions must be made, one of which concerns the approach needed to account for the time variation of the applied magnetic fields. A similar point arises in at least two of the other theories discussed later in this paper. Eventually the theory is presented as a series of plots used to show how the ligand velocity is modified by the parameters in the complex theoretical expression. The results suggest that the amplitude of the AC magnetic field could play a key roll in controlling the biological endpoint. Conditions are shown to exist where ligand velocity maxima can occur near integer multiples (harmonics) of the cyclotron resonance frequency. There does not appear to be any preference between even and odd integers, and there is no evidence of maxima not occurring at certain harmonic values. This model can be rejected, since it is hard to envision how even harmonics could lead to a net energy transfer.

A model that is elegant in the fact that it consists of the simple essentials has been presented by Durney et al. (1988). Their system consists of a single charged particle moving in a viscous medium. The particle is subjected to an AC and a DC magnetic field and an AC electrical field induced by the AC magnetic field. The basic equation of motion (Lorentz force equation) is used to start the analysis, and a set of simultaneous scaler differential equations result. Many of the coefficients in the equations can be written in the form of Equation 1. As was pointed out earlier, this is a result of beginning with the Lorentz equation. This set of linear differential equations describing the simple model does not have a closed form solution because the coefficients of the equations vary with time. Chiabrera and Bianco (1987) faced a similar difficulty in their model. Durney's group (1988) proceeded to solve the equations of motion through the use of numerical techniques, then presented a set of remarkable plots showing the response of the charged particle as several parameters of the driving fields were varied. The model did predict both amplitude and frequency windows, but only if the viscous damping in the medium was very low. This is consistent with the first model discussed wherein ligand coupling to the ELF fields was considered possible only in the microenvironment of the binding site. The Durney model suggests rather conclusively that a charged particle moving in a viscous solution will not exhibit a resonant response at ELF frequencies. As pointed out by Durney et al.

(1988), the model is difficult to extrapolate to a biological system except in the most general sense. This model, like the Chiabrera model, predicts resonances can occur even if the DC magnetic field is zero. This is directly contradicted by experimental evidence that shows the biological response does not occur if either the AC field or the DC field is set to zero (Carson et al., 1990; Liboff et al., 1987, 1990; McLeod et al., 1987; Smith et al., 1987; Walheczek and Liburdy, 1990). As with the previous model, there is no pattern in the harmonics where resonances appear. The model does not address the thermal energy (kT) question since it is not suggested as a theory that can be extrapolated to a biological system.

A new model was proposed by Lednev (1990). In this case, the ion is considered to be surrounded by a hydration shell or bound in a calcium-binding protein to oxygen ligands. It is then stated that the ion may be considered as a charged spatial oscillator with a characteristic set of vibrational frequencies. Thermal motion may cause coordination bonds to form and break continuously, which would imply a complex set of energy levels in which the ion could exist. The analysis proceeds by assuming the ion has only one excited level. Without this assumption, the complexity of the situation would preclude any possible mathematical description (a pertinent point to consider with any of the proposed theories is the extent of limitation imposed on the final results by the assumptions that are made for the theory to proceed). Lednev develops the probability that the ion will make the transition to the energy ground state and shows that the probability increases to a maximum at multiples of the cyclotron resonance frequency. The conclusion is then drawn that weak magnetic fields will alter the state of equilibrium of the ion and its surroundings, which in turn may alter the biological state. This theory addresses the thermal energy (kT) question by suggesting that thermal motion is a necessary part of this model, since such motion is assumed to drive the ion energy quantizing. This seems to violate the rules of energy quantization. To get a change in energy level requires one quantum of energy (either energy in or out), no more, no less. The kT energy is not necessarily equal to any particular quantum energy level, and it cannot be accumulated to cause an energy level transition.

Lednev also assumed that the ion oscillation frequencies are determined only by interactions with six ligand "nearest neighbors," so he could write a set of energy level exclusion rules. This led to the prediction that only harmonics 1, 3, 5, and 15 of the ELF field would be effective in maximizing the transition probability. Lednev suggests that this model fits much of the experimental data and, in particular, finesses the thermal problem since the weak ELF fields at the CR frequency act only to increase the probability of the ion to transition into a ground energy state. It does not seem, however, that the latter statement really answers the (kT) question, as pointed out above. The theory is, however, a very interesting new approach to the modeling of the ELF biological system coupling mechanism.

Liboff (1985) observed that cyclotron resonance could explain why Blackman's data (Blackman et al., 1979, 1982, 1985a, 1985b) changed as a function of the DC magnetic field. The suggestion was received by the research community with mild interest since the ion resonance frequencies for a number of biologically important ions are below 100 Hz for a DC magnetic field in the range of the Earth's geomagnetic field 41 (about 50 μ T). Two papers then developed a simple model (McLeod and Liboff, 1986; Liboff and McLeod, 1987) suggesting that although the ions in a biological solution would not experience any resonance [as pointed out by Durney et al. (1988)], it could be possible that a low-collision volume might exist in the membrane channel through which the ions pass. If the ELF fields couple to an ion such as Ca^{2+} and subsequently modify its rate of flow through the channel, a biological change in the cell would follow. This model suggested that the solvation shell of the ion was not involved while the ion was in the channel (somewhat like the microenvironment argument in Chiabrera's model). This microenvironment involving loss of the hydration shell and low collisions we call Liboff Space. Many arguments were raised about why the CR model could not be correct (reviewed by Liboff, 1990; Liboff et al., 1989), but two specific experiments were reported in 1987 (Rozek et al., 1987; McLeod et al., 1989a, 1989b) that supported the model.

The simplistic theory was later expended by McLeod et al., 1989a, 1989b; McLeod and Liboff, 1992; Toyoshima and Unwin, 1988) to incorporate some of the latest results on the interior shape of certain biological channels. Toyoshima and Unwin (1988) developed a closed form result for the ion movement that suggests that for a specific channel cross-section [such as the acetylcholine receptor channel reported by Toshiema and Unwin (1988)] the ELF magnetic fields can cause the ion movement to "fit" the channel at the CR frequency. It is also shown that the harmonics 1, 3, 5, and 15 will cause the ion path to fit the channel, whereas no other harmonics will cause a fit. A vital assumption in the development of this model is that the ion moves through the channel in roughly the necessary time to complete one cycle of the ELF signal. The assumed transit time is very slow when compared to ion transit times measured for action potentials or in most voltage-controlled channels. It is suggested in this model that there may be slow channels that activate only in response to an attempt by the cell to correct a non-equilibrium situation.

This theory can be criticized for insufficient modeling of the forces inside the channel and because it has to postulate the existence of a slow channel if the model is to work. It is interesting, however, that the exclusion of all even harmonics and all but the 1st, 3rd, 5th, and 15th harmonics does emerge in the theory. The theory attempts to deal with the (kT) question from the narrow-banding (as opposed to the time-averaging) standpoint. The channel shape acts as a filter to allow only a narrow range of frequencies to couple to the ion effectively.

Magnetic Quantum Number and the Zeeman Effect

The model used by Weller et al. (1984) was an application of the Bohr magneton when the total quantum number breaks up into several sub-states when the atom is in a magnetic field and the energies are slightly more or slightly less than the energy of the state in the absence of the field. Weller studied the production of radical-pair combinations. Under the effect of a weak magnetic field, the spin realignment of the initial radical ion pair in a chemical reaction is governed by the hyperfine reaction between the unpaired electron spin and nuclear spins. The magnetic field can influence the exchange interaction of the radical spins in the pair. A perturbation energy of 10^{-7} eV for a few nanoseconds was experimentally confirmed. Shulten (1986) extended this work to implications for biological systems (Digby 1966). One slow-acting and key system could be the regulatory G proteins, which couple an external harmonal signal to the intracellular effector mechanisms (Jones et al., 1987). Thus, two targets could be considered, one an electron transfer reaction in an enzyme and the other the interaction of an ion (calcium) with its binding site.

Quantum Wave Resonance Model

In this model, the site of ELF action on calcium is not thought to be at the membrane, where the time constants are too low (1 msec), but on protein-binding sites for calcium, as in the Lednev model, where the binding constants might be in the order of seconds.

Calcium is a good target compared to other ions for biological effects because small concentration changes, which represent a small number of ions (about 6000), can activate many cellular processes. Other ions do not play such an important regulatory role (if the cell received 60,000 potassium ions, for instance, it probably would not be noticeable). It is thought that although calcium binds to calmodulin for about 1 msec in free solution, the calcium–calmodulin–target protein interaction might be much slower (up to 1 sec). Inside the calcium-binding site on calmodulin there is no water and, hence, no hydration shell. A single water molecule bridges the opening in the binding site.

If, in the Liboff Space mentioned above, the calcium ion is viewed as a wave function that describes the ion's interactions with the boundaries of the space and if this wave function is modulated by the resonating magnetic fields, these fields could change the affinity of the calcium ion for its target.

DISCUSSION

It appears that the effects reported do not all sit happily together in one hypothesis. Galvanotaxis in a DC field or AC electrical field of necessity requires positional information to be given to the cell. The CR type of theories do not explain this. Static magnetic field interactions are also not explainable by CR theories but perhaps by quantum mechanical effects. Experiments need to be conducted to more closely study the difference between the induced electrical field and the magnetic field component. Evidence of quantum mechanical effects involving magnetic quantum number phenomena in DC magnetic fields at very short time scales and very low field strengths do not preclude other quantum mechanical effects due to field resonances. Further the implication of the measurements for integrated biological systems have not been explored. Is it then important to control for the static Earth's magnetic field during electrical field and PEMF experiments? It would seem so, if only to preclude interference from another interaction of the same order of magnitude. Perhaps, too, normal biological experiments should be conducted in wholly controlled environments if the literature is to be only partly believed.

A person attempting to evaluate the state of the research in defining an acceptable model for ELF biological system interaction is probably going to develop a sense of frustration after his or her first look at these diverse theories. A legitimate question must now be addressed. How does one begin to test the validity of each or all of the above models? It does little good at this point for theoreticians to criticize or endorse any of the models. Each has some valid points, and each has some weaknesses. Perhaps one straightforward way to test the theories would be to set up experiments according to the prediction parameters of each model, then to measure the time it takes the biological system to show a response. At the very least, this would suggest whether signal averaging or narrow banding was operating within the biosystem. The use of channel blockers or binding site competitors in conjunction with the ELF fields could perhaps pinpoint whether the membrane surface or the channel through the membrane is the site of the interaction. There are several new techniques available for measuring the flow of ions, such as calcium, into the cell interior. If real time experiments can be designed using these techniques, one could look for the movement of the Ca^{2+} ion across the cell membrane as the ELF field is applied. If a biological response occurs with no measurable ion flow into the cytoplasm, it would suggest that the ELF field is coupling to the surface of the membrane rather than to an ion moving across the membrane in a channel.

A final, but vital point of discussion, is the question of the biological system that could be used to test the theories. There are as many biosystems in use as there are experiments being done, it seems. This is reasonable since each research laboratory has people that are expert in certain measurement techniques and certain biological systems. The problem is that given the wide diversity of laboratories working on defining a mechanism of interaction, it is nearly impossible to repeat and hence verify the research results of any one laboratory. Protocols are often exchanged between laboratories and "identical" experiments are performed, with the result being different experimental data. Careful investigation usually reveals that some minor variations existed in the execution of the protocol. Unfortunately, it is not possible to tell at this point just what defines a "minor variation." It seems clear that a single biological system should be selected, and experiments designed to answer single questions should be designated. If an experimental protocol were then written and executed exactly by three to five separate laboratories, some of the answers to the questions posed here might emerge. It appears that it is time for researchers to step back from doing more experiments and to work together to select the biosystem, the experiment, and the protocol that has a reasonable chance to produce results that can lead to a mechanism of interaction between ELF fields and biological systems.

SUMMARY

A review of most current theories on how weak electromagnetic fields might affect cells fails to identify any plausible mechanism. Many biological experiments appear to be flawed in one way or another and are generally unreproducible, either between laboratories or within a single laboratory. This lack of concensus is suggestive of an artifactual source for the reported data.

References

Adey, W. R. (1981). Ionic non-equilibrium phenomena in tissue interactions with nonionizing electromagnetic fields. Page 271. In Biological Effects of Nonionizing Radiation. Illinger, K. H., ed. Washington, American Chemical Society, Symposium Ser. No. 157.

Adey, W. R. (1988). Physiological signalling access cell membranes and cooperative influences of extremely low-frequency electromagnetic fields. Pages 148–170. In Biological Coherence and Response to External Stimuli. Fröhlich, H., ed. Springer-Verlag, Berlin.

Adey, W. R. (1990a). Nonlinear electrodynamics in cell membrane transductive coupling. Pages 1–27. In Membrane Transport and Information Storage, Alan R. Liss, New York.

Adey, W. R. (1990b). Electromagnetic fields and the essence of living systems. Pages 1–36. In Modern Radio Science. Andersen, J. B., ed. Oxford University Press, Oxford.

Arendse, M. C. (1978). Magnetic field detection is distinct from light detection in the invertebrates Tenebrio and Talitrus. Nature **24**: 358–362.

Barker, A. T., Dixon, R. A., Sharrard, W. J. W., and Sutcliffe, M. L. (1984). Pulsed magneticfield therapy for tibial non-union. The Lancet **May 5**, 994–996.

Baker, R. R., Mather, J. G., and Kennaugh, J. H. (1983). Magnetic bones in human sinuses. Nature **301**: 78–90.

Bassett, C. A. L., and Becker, R. O. (1962). Generation of electric potentials by bone in response to mechanical stress. Science **137**: 1063–1064.

Bassett, C. A. L., Choksi, H. R., Hernandez, E., Pawluk, R. J., and Strop, M. (1979a). The effect of pulsing electromagnetic fields on cellular calcium and calcification of non-unions. In Electrical Properties of Bone and Cartilage. Brighton, C. T., Black, J., and Pollack, S. R., eds. Grune & Stratton, New York.

Bassett, C. A. L., and Hess, K. (1984) Synergistic effects of PEMF and fresh canine cancellons bone grafts. Trans. O.R.S. **30**: 49.

Bassett, C. A. L., Mitchell, J. N., and Gaston, S. R. (1981). Treatment of ununited tibial diaphyseal fractures with pulsing electromagnetic fields. J. Bone and Joint Surgery **36-A**: 511–523.

Bassett, C., Mitchell, S., and Gaston, S. (1982). Pulsing electromagnetic treatment in ununited fractures and failed arthrodeses, J. Am. Med. Assoc. **247**: 623.

Bassett, C. A. L., Pawluk, R. J., and Pilla, A. A. (1974). Argumentation of bone repair by inductively coupled electromagnetic fields. Science **184**: 575–577.

Bassett, L. S., Tzitzikalakis, R. S., Pawluk, R. J., Bassett, C. A. L. (1979b). Presentation of disuse osteoporosis in the rut by means pulsing magnetic fields. Pages 311–331. In Electrical Properties of Bone Cartilage: Experimental Effects and Clinical Applications. Brighton, C. T., Block, J., and Pollack, J. R., eds. Grune & Stratton, New York.

Bawin, S. M., and Adey, W. R. (1976). Sensitivity of calcium binding in cerebral tissue to weak electric fields oscillating at low frequency. Proc. Natl. Acad. Sci. USA **73**: 1999.

Bawin, S. M., Adey, W. R., and Sabbot, I. M. (1978b). Ionic Factors in release of 45 Ca^{2+} from chick cerebral tissue by electromagnetic fields. Proc. Natl. Acad. Sci. USA **75**: 6314.

Bawin, S. M., Kaczmarek, L. K., and Adey, W. R. (1975). Effects of modulated VHF fields on the central nervous system. Ann. N.Y. Acad. Sci. **247**: 74.

Bawin, S. M., Sheppard, A. R., and Adey, W. R. (1978a). Possible mechanisms of weak electromagnetic field coupling in brain tissue. Bioelectrochem. Bioenerg. **45**: 67.

Beischer, D. E. (1968). Biomagnetics. Ann. N.Y. Acad. Sci. **134**: 454–468.

Binder, A., Paw, G., Hazleman, B., and Fitton-Jackson, S. (1984). Pulsed electromagneticfield therapy of persistent rotator cuff tendonitis a double blind controlled Assessment. The Lancet, **March 31**.

Bioelectromagnetics Society Newsletter (1990). September/October, **96**: 8–9.

Blackman, C. F., Benane, S. G., House, D. E., and Joines, W. T. (1985a). Effects of ELF (1-120HZ) and modulated (50 Hz) RF fields on the efflux of calcium ions from brain tissue, in vitro. Bioelectromagnetics **6**: 1.

Blackman, C. F., Benane, S. G., Kinney, L. S., Joines, W. T., and House, D. E. (1982). Effects of ELF fields on calcium ion efflux from brain tissue, in vitro. Radiat. Res. **92**: 510.

Blackman, C. F., Benane, S. G., Rabinowitz, J. R., House, D. E., and Joines, W. T. (1985b). A role for the magnetic field in the radiation-induced efflux of calcium ions from brain tissue in vitro. Bioelectromagnetics. **6**: 327.

Blackman, C. F., Elder, J. A., Weil, C. M., Benane, S. G., Eichinger, D. C., and House, D. E. (1979). Induction of calcium ion efflux from brain tissue by radio frequency radiation. Radio Sci. **14**: 93.

Blakemore, R. P. (1975). Magnetotactic bacteria. Science **190**: 377–379.

Blank, M. (1982). The surface compartment model (SCM): Role of surface charge in membrane permeability changes. Bioelectrochem. Bioenerg. **9**: 615–624.

Blank, M. (1983). The surface compartment model (SCM) with a voltage sensitive channel. Bioelectrochem. Bioenerg. **10**: 451–465.

Blank, M. (1984). Properties of ion channels inferred from the surface compartment model (SCM). Bioelectrochem. Bioenerg. **13**: 93–101.

Blank, M. (1987). Ionic processes at membrane surfaces: The role of electrical double layers in electrically stimulated ion transport. Pages 1–13. In Mechanistic Approaches to Interactions of Electric and Electromagnetic Fields with Living Systems. Blank, M., and Findl, E., eds. Plenum Press, New York.

Blank, M., and Kavanaugh, W. P. (1982). The surface compartment model (SCM) during transients. Bioelectrochem. Bioenerg. **9**: 427–438.

Blank, M., Kavanaugh, W. P., and Cerf, G. (1982). The surface compartment model: Voltage clamp. Bioelectrochem. Bioenerg. **9**: 439–458.

British Medical Journal Secretariat (1980). Editorial: Electricity in Bones. Brit. Med. J. 16th August, 1980.

Bruce, G. K., Howlett, C. R., and Huckstep, R. L. (1987). Effect of a static magnetic field on fracture healing in a rabbit radius. Clin. Orthop. Rel. Res. **222**: 300–306.

Byus, C. V., Lundak, R. L., Fletcher, R. M., and Adey, W. R. (1984). Alterations in protein kinase activity following exposure of cultured human lymphocytes to modulated microwave fields. Bioelectromagnetics **5**: 341–351.

Carson, J. J. L., Prato, F. S, Drost, D. J., Diesborg, L. D., and Dixon, S. J. (1990). Time-varying magnetic fields increase cytosolic free Ca^{2+} in HL60 cells. Paper E-2-4, Page 48. Abstracts Twelfth Annual Meeting, Bioelectromagnetics Society, San Antonio, TX.

Chiabrera, A., and Bianco, B. (1987). The role of the magnetic field in the EM interaction with the ligand binding. Pages 79–95. In Mechanistic Approaches to Interactions of Electric and Electromagnetic Fields with Living Systems. Blank, M., and Findl, E., eds. Plenum Press, New York.

Conley, C. C. (1969). Effects of near-zero magnetic fields upon biological systems. In Biological Effects of Magnetic Fields, vol. 2. Barnothy, M. F., ed. Plenum Press, New York.

Cooper, M. S. (1984). Gap junctions increase the sensitivity of tissue cells to exogenous electric fields. J. Theor. Biol. **111**: 123–130.

Digby, P. S. B. (1966). Mechanism of calcification in mammalian bone. Nature **212**: 1250–1252.

Dixey, R., and Rein, G. (1982). 3H noradrenaline release potention in a clonal nerve cell by low intensity pulsed magnetic fields. Nature **296**: 253–256.

Durney, C. H., Rushforth, C. K., and Anderson, A. A. (1988). Resonant AC–DC magnetic fields: Calculated response. Bioelectromagnetics 9: 315–336.

Farndale, R. W., and Murray, J. C. (1985). Pulsed electromagnetic fields promote collagen production in bone marrow fibroblasts via athermal mechanisms. Calcif. Tissue Int. 37: 178–182.

Fitzsimmons, R. J., Farley, J., Adey, W. R., Baylink, D. J. (1986). Embryonic bone matrix formation is increased after exposure to a low amplitude capacitively coupled electric field in vitro. Biochim. Biophys. Acta 882: 51–56.

Foster, K. R., and Pickard, W. F. (1987). Microwaves: The risks of risk research. Nature 330: 531–532.

Fukada, E., and Yasuda, I. (1957). On the piezoelectric effect in bone. J. Phys. Soc. Jpn. 12: 1158.

Goodman, R., Bassett, C. A. L., and Henderson, A. S. (1983). Pulsing electromagnetic fields induce cellular transcription. Science 220: 1283–1285.

Goodman, R., and Henderson, A. (1986a). Some biological effects of electromagnetic fields. Bioelectrochem. Bioenerget. 15: 39.

Goodman, R., and Henderson, A. (1986b). Sine waves induce cellular transcription. Bioelectromagnetics 7: 23.

Goodman, R., and Henderson, A. (1987a). Transcriptional patterns in the X-chromosome of Sciara coprophila following exposure to magnetic fields. Bioelectromagnetics 8: 1.

Goodman, R., and Henderson, A. (1987b). Patterns of transcription and translation in cells exposed to EM fields: A review. Page 217. In Mechanistic Approaches to Interactions of Electric and Electromagnetic Fields with Living Systems. Blank, M., and Findl, E., eds. Plenum, New York.

Jaffe, L., and Poo, M. M. (1979). Neurites grow faster towards the cathode than the anode in a steady field. J. Exp. Zool. 209: 115–117.

Jones, D. B. (1984). The Effect of PEMF on cAMP metabolism in cultured chick embryo tibiae. J. Bioelectricity 3: 427–451.

Jones, D. B., Bolwell, G. P., and Gilliatt, G. (1986a). Amplification, by PEMF, of plant growth regulator induced phenylalanine ammonialyase during differentiation in suspension cultured plant cells. J. Bioelectricity 5: 1–12.

Jones, D. B., Pedley, R. B., and Ryaby, J. T. (1986b). The Effects of PEMF on differentiation and growth in cloudman S91 murine melanoma cells in vitro. J. Bioelectricity 5: 145–169.

Jones, D. B., and Ryaby, J. T. (1987). Low-energy, time-varying electromagnetic field interactions with cellular control mechanisms. In Mechanistic Approaches to Interactions of Electric and Electromagnetic Fields with Living Systems. Blank, M., and Findl, E., eds. Plenum Press, New York.

Jones, D. B., Ryaby, J. T., and Pilla, A. A. (1987). GTP-binding proteins may be the regulatory site of interaction of PEMF in melanoma cells. J. Orthop. Res. Trans. BRAGS 6: 45.

Law, H. T., Annan, I. D., Hugmes, S. P. F., Stead, A. C., Camburn, M. A., and Montgomery, H. (1985). The effect of induced electric currents on bone after experimental osteotomy in sheep. J. B. J. Surg. 67B 3: 463–469.

Lawrence, A. F., McDaniel, J. C., Chang, D. B., and Birge, R. R. (1987). The nature of phonons and solitary waves in alpha-helical proteins. Biophys. J. 51: 785.

Lednev, L. L. (1990). Possible mechanism for the influence of weak magnetic fields on biological systems. Bioelectromagnetics 12: 71–75.

Liboff, A. R. (1985). Cyclotron resonance in membrane transport. In Interactions

Between Electromagnetic Fields and Cells. Page 281. Chibrera, A., Nicolini, C., and Schwan, H. P., eds. Plenum, London.

Liboff, A. R. (1990). Interaction mechanisms of low-level electromagnetic fields in living systems. Ramel, C., and Norden, B., eds. Oxford Press, Oxford.

Liboff, A. R., and McLeod, B. R. (1987). Kinetics of channelized membrane ions in magnetic fields. Bioelectromagnetics 9: 39–51.

Liboff, A. R., McLeod, B. R., and Smith, S. D. (1989). Ion cyclotron resonance effects of ELF fields in biological systems. Page 251. In Extremely Low Frequency Fields: The Question of Cancer. Wilson, B. W., Stevens, R. G., and Anderson, L. E., eds. Battelle Press, Columbus, OH.

Liboff, A. R., McLeod, B. R., and Smith, S. D. (1990). Ion cyclotron resonance effects of ELF fields in biological systems. In Extremely Low Frequency Electromagnetic Fields: The Question of Cancer. Pages 251–290. Wilson, W. B., Stevens, R. G., and Anderson, L. E., eds. Baltelle Press, Columbus, OH.

Liboff, A. R., Smith, S. D., and McLeod, B. R. (1987). Experimental evidence for ion cyclotron resonance mediation of membrane transport. Page 109. In Mechanistic Approaches to Interactions of Electromagnetic Fields with Living Systems. Blank, M., and Findl, E., eds. Plenum Press, New York.

Luben, R. A., Cain, C. D., Chen, C-Y., Rosen, D. M., and Adey, W. R. (1982). Effects of electromagnetic stimulation on bone and bone cells in vitro: Inhibition of response to parathyroid hormone by low energy low frequency fields. Proc. Natl. Acad. Sci. USA 79: 4180–4148.

Lyle, D. B., Ayotte, R. D., Wang, Z., Sheppard, A. R., and Adey, W. R. (1989). Activation and proliferation of normal and leukemic T-Lymphocytes exposed to magnetic fields under calcium cyclotron resonance conditions. Paper B-3-3, Page 13. Abstracts Eleventh Annual Meeting, Bioelectromagnetics Society, Tucson, AZ.

McLeod, B. R., and Liboff, A. R. (1986). Dynamical characteristics of membrane ions in multi-field configurations at low frequencies. Bioelectromagnetics 7: 177–189.

McLeod, B. R., and Liboff, A. R. (1992). Electromagnetic gating in ion channels. J. Theor. Biol. 158: 15–31.

McLeod, B. R., Liboff, A. R., and Smith, S. D. (1989a). A theoretical model that predicts frequency, amplitude and harmonic resonances in cell channels. Paper E-4-3. Eleventh Annual Meeting of the Bioelectromagnetics Society, Tucson, AZ.

McLeod, B. R., Liboff, A. R., and Smith, S. D. (1989b). A mathematical model incorporating membrane channel parameters that exhibits frequency, amplitude and harmonic windows. Page 51. Ninth Annual Meeting of the Bioelectrical Repair and Growth Society, Cleveland, OH.

McLeod, B. R., Smith, S. D., and Liboff, A. R. (1987). Ion cyclotron resonance frequencies enhance Ca^{2+} dependent motility in diatoms. J. Bioelectricity 6: 1–12.

Mulier, J. C., and Spaas, F. (1980). Out-patient treatment of surgically resistant non-unions by induced pulsing current-clinical results. Arch. Orthop. Traumat. Surg. 97: 293–29.

Murray, J. C., and Farndale, R. W. (1985). Modulation of collagen production in cultured fibroblasts by a low-frequency pulsed magnetic field. Biochim. Biophys. Acta 838: 98–105.

Raybourn, M. S. (1983). The effects of direct-current magnetic fields on Turtle retinas in vitro. Science 220: 715–717.

Reuss, S., and Olcese, J. (1986). Magnetic field effects on the rat pineal gland: Role of retinal activation by light. Neurosci. Lett. 64: 97–101.

Reuss, S., Semm, P., and Vollrath, L. (1984). Different types of magnetically sensitive cells in the rat pineal gland. Neurosci. Lett. 40: 23–26.

Rooze, M., and Hinsenkamp, M. (1985). In vivo modifications induced by electromagnetic stimulation of chicken embryos. Reconstr. Surg. Traumat. 19: 87–92.

Ross, S. M. (1988). Effects of sinusoidal electromagnetic fields on proliferation of rabbit ligament fibroblasts. Page 68. Translations of the Eighth Annual Meeting, Bioelectrical Repair and Growth Society, Washington, D.C.

Rozek, R. J., Sherman, M. L., Liboff, A. R., McLeod, B. R., and Smith, S. D. (1987). Nifedipine is an antogonist to cyclotron resonance enhancement of −45Ca incorporation in human lymphocytes. Cell Calcium 8: 413–427.

Shulten, K. (1982). Magnetic field effects in chemistry and biology. Festkörperprobleme 22: 61–83.

Shulten, K. (1986). Magnetic field effects in chemical and biological photoprocesses. Proc. Workshop on Biophysical Effects of Steady Magnetic Fields, Les Houches. Springer-Verlag, Heidelberg.

Shulten, K., and Wolynes, P. G. (1978). J. Chem. Phys. 68: 3292–3295.

Sisken, B. F., and Smith, S. D. (1975). The effects of minute direct electrical currents on cultured chick embryo trigeminal ganglia. J. Embryol. Exp. Morphol. 33: 29–34.

Smith, S. D., McLeod, B. R., Liboff, A. R., and Cooksey, K. E. (1987). Calcium cyclotron resonance and diatom motility. Bioelectromagnetics 8: 215–227.

Smith, R. L., and Nagel, D. A. (1983). Effects of pulsing electromagnetic fields on bone growth and articular cartilage. Clin. Orthop. Rel. Res. 181: 277–282.

Thurm, U. (1983). Mechano-electric transduction. Pages 666–671. In Biophysics. Hoppe, W., Lohmann, W., Markl, H., and Ziegler, H., eds. Springer Verlag, Berlin, Heidelberg.

Toyoshima, C., and Unwin, N. (1988). Ion channel of acetylcholine receptor reconstructed from images of postsynaptic membranes. Nature 336: 247.

Walheczek, J., and Liburdy, R. P. (1990). Combined DC/AC magnetic fields alter Ca^{2+} metabolism in activated rat thymic lymphocytes, Paper D-2-1, Page 39. Abstracts Twelfth Annual Meeting, Bioelectromagnetics Society, San Antonio, TX.

Weaver, J. C., and Astumian, R. D. (1990). The response of living cells to very weak electric fields: The thermal noise limit. Science 247: 459–462.

Weller, A., Staerk, H., and Treichel, R. (1984). Magnetic-field effects on geminate radical-pair recombination. Farday Discuss. Chem. Soc. 78: 271–278.

Yamada, S., Guenther, H. L., and Fleisch, H. (1985). The effect of PEMF on bone cell metabolism and cavaria resorption in vitro and on calcium metabolism in the live rat. Int. Orthopaedics 9: 129–134.

CHAPTER 12

Electrostimulation of Cells by Weak Electric Currents

A. Goldsworthy

ABSTRACT

Some cells can detect voltage gradients as low as 0.5 $\mu V / m$ and current densities of as little as 5 nA / cm^2. These extremely sensitive electrosensing mechanisms are used by animals for navigation and to find prey and by plants for predicting the availability of water. They probably evolved from somewhat less-sensitive mechanisms used by most cells for sensing the weak currents that flow within organisms to control their growth. The artificial application of weak electrical currents can sometimes interact with these to stimulate growth, the healing of injuries, and the regeneration of organs. Examples are given, the mechanisms of the effects discussed, and a hypothesis presented for their likely evolution.

INTRODUCTION

Reports of weak electrical currents stimulating growth and organ regeneration date back more than 200 years. The effects occur in plants and animals and also in their cultured tissues. There is a wide and disparate literature on the subject and a number of reviews covering different aspects; for example, Briggs et al. (1926); Ellis and Turner (1978); Quatrano (1978); Nuccitelli (1984); Schnepf (1986); and Borgens et al. (1989). It is beyond the scope of this chapter to write a comprehensive review of the whole field. Instead, some of the more interesting and important findings are discussed and attempts made to relate them to a common theme.

ELECTROSTIMULATION OF PLANTS

Experiments on the electrostimulation of plants date back to Maimbray in 1746 who studied the effects of electricity on myrtle bushes, but most of the early experiments were on a very small scale and frequently gave contradictory results. Lemstrom (a professor of physics at Helsinki) was the first to make a large-scale investigation of the apparently beneficial effects of weak electrical currents on plants. He paid several visits to the polar regions and was impressed by the relatively lush growth of vegetation, despite the low temperature and poor light. He noted that fir trees growing there showed a periodicity in their growth rings corresponding to the sunspot cycle and suspected that there may be a stimulation of growth due to weak electrical currents flowing through the atmosphere from the aurora borealis. He then set up experiments to test this by applying artificial electrical fields from an electrostatic generator via overhead wires to a wide range of crops. The voltages were not quoted directly but were usually sufficient to cause a spark of 1–2 mm between charged balls (about 3–6 kV), giving gradients in the region of 7–14 kV/m between the wires and the crop. He conducted many experiments over several years in a number of European countries and found an average yield stimulation of about 45%. The results, however, were not always consistent and appeared to depend upon the weather. In particular, the application of electrical fields during hot dry weather was injurious and could even decrease yield.

Lemstrom's (1904) findings led to a flurry of activity by other workers seeking to exploit the effect over the next few decades, the technique being termed *electroculture*. Their efforts, however, met with only mixed success. In some cases, spectacular stimulations were found. For example, Blackman (1924), working at Rothamsted in the United Kingdom, grew crops under a wire mesh about 2 m above ground level that was charged for 6 hours a day, giving gradients to Earth of 20–40 kV/m. The current (which flows to the plants via air ions) was about 0.5–1 mA per acre (about 50–100 pA per plant). Within 20 days, treated barley plants were taller and greener than the controls. Out of a total of 18 trials with spring wheat, barley, and oats, 14 gave significant increases in dry weight, 9 of which were more than 30%. In a parallel series of pot experiments (Blackman and Legg 1924), it was found that currents from 10 pA to 10 nA per plant promoted growth, but higher values were injurious. Direct currents of either polarity were equally effective, but the best results were obtained with alternating currents. Prolonged exposure seemed to be unnecessary, since electrical treatment for only the first month gave the same response as exposure for the whole growing period. In fact, even shorter periods may be sufficient. Blackman et al. (1923) measured the growth rate of barley coleoptiles at 15-min intervals during and after exposure to the current. The growth rate (compared to untreated controls) began to increase within an hour and continued to increase for at least 4 hours after the current was switched off. This indicates that the effect of electrostimulation was to trigger a process leading to more rapid growth, which was then self-sustaining.

Even seeds have been reported to respond to electrostimulation. Lutkova and Oleshko (1965) found that a current of 0.57 mA/cm^2 accelerated the breaking of dormancy when applied to moist cherry seeds during stratification. They did not test any lower currents, but Shatilov and Trifonova (1968) found significant stimulation of barley seed germination even with currents as low as 15 nA/cm^2 applied for only 5 min. In both cases, the stimulations were associated with an increased mobilization of food reserves and greater enzyme activity. Once the electrical treatment has been given, the effect seems stable. For example, Stanko and Koshevnikova (1972) reported a stimulation of the mobilization of food reserves in wheat seed after being placed in high-voltage fields (100–800 kV/m for 30 min), even when the seeds were stored for up to 38 days between treatment and germination. A similar effect was reported by Pittman and Ormrod (1970) when they exposed wheat seeds to a magnetic field. Perhaps the stimulatory effect in this case was due to a displacement of charge or weak currents generated within the seeds as they were moved into and out of the field.

Not all workers found similar effects. For example, Ellis (1981) failed to find stimulation of the germination of barley seed exposed to electrostatic fields, and Briggs et al. (1926), working in the United States, were unable to obtain significant stimulation of crop growth, despite many years of trying. Even Blackman (1924), working in the United Kingdom, was unsuccessful in all his attempts, and, in some cases, electrical treatment led to a reduction in yield. These discrepancies were ascribed to differences in weather conditions and the different climates in the United States and the United Kingdom. Any rational explanation of these growth stimulations must therefore include a plausible explanation for why they are not always reproducible and how they might be affected by the weather. It should also include some indication of what advantage electrostimulation could be to the plant to explain how it might have evolved by natural selection.

Natural Electrostimulation

If the artificial electrical stimulation of growth is beneficial, why have plants and seeds not evolved to grow equally fast without it? A possible answer is that plants have evolved to use the "natural" electrical currents that flow through the atmosphere as cues from the environment to stimulate their growth when conditions are suitable. If so, current applied under the wrong conditions may be detrimental, and its indiscriminate application may yield non-reproducible results.

Natural Atmospheric Currents

If the ability to respond to electrostimulation arose by natural selection, similar voltages and currents would be expected to occur naturally and the ability to respond confers some selective advantage. Natural currents normally flow vertically through the atmosphere with strengths that depend on the weather (Chalmers 1957; Malan 1963). In fine weather there is a modest voltage gradient of about 150–300 V/m near the Earth's surface (the Earth is negative), giving current densities of about 1 pA/m^2. In wet weather, the voltage gradient increases and may also reverse in direction. In a thunderstorm, the gradients increase dramatically until there is a lightning flash. Even before the lightning, however, quite large currents flow to Earth by point discharge (a localized discharge of ions around sharp points such as the tips of leaves and spines on plants). Schonland (1928) measured voltage gradients of up to 16 kV/m under thunderclouds, which is within the range reported to cause the electrostimulation of growth. He also measured the current flowing through a small acacia tree. He placed the tree on insulators, connected it to Earth through a galvanometer, and found values up to 4 μA at the center of a thunderstorm. He estimated that the tree was the main center for point discharge for the surrounding 20 m^2, giving a mean current density of 0.2 μA/m^2 flowing to Earth. This is approximately 0.8 mA/acre and within the range routinely used by Blackman (1924) and others to produce electrostimulation. Thus, voltage gradients and currents that are sufficient to cause the electrostimulation of plant growth occur naturally during thunderstorms, and probably the lush growth of vegetation often seen after such a storm (which no amount of treatment with the watering can or garden hose seems to mimic) may have as much to do with electrostimulation as with the rain.

The Selective Advantage

The advantage of natural electrostimulation may be that it enables plants and dormant seeds to predict the imminent availability of large amounts of water. In adult plants, this would allow them to put the machinery of protein synthesis and growth into operation (a process taking several hours) in time to take full advantage of the rainfall. If they had to wait until the water percolated to root level before responding, hours of potential growing time would be lost. In the case of dormant seeds, it may be even more important. Because of their hydrophillic nature, seeds can become fully imbibed, even when the soil moisture is insufficient to support seedling growth. Such seeds can remain dormant for years until they receive a signal from the environment indicating that conditions are suitable for germination. In many cases, this happens when heavy rain washes inhibitors out of the seed, but natural electrostimulation from weak currents flowing through the soil in thunderstorm conditions could be an even better signal. It would predict the imminent availability of sufficient water for seedling establishment and initiate germination in time to take the greatest advantage from it.

Why Electrostimulation is not Always Successful

In some cases, the failure to obtain stimulations might be due to the over-enthusiastic application of current. For example, Murr (1963) tested voltage gradients of 12–70 kV/m applied continuously to orchard grass seedlings. Although the treated plants became greener than the controls, there was no stimulation of growth and there were visible signs of burning (especially at the higher voltages) around pointed regions of the plant (which are the main locations of current entry). Even so, this does not explain why the milder treatments at lower voltages or for shorter periods given by other workers were sometimes effective and sometimes not.

   The hypothesis that natural electrical fields are beneficial because they stimulate growth before the arrival of water from heavy rain does, however, explain why indiscriminate artificial electrostimulation is not always successful. If it is given when there is insufficient water to support the extra growth, it is likely to prove injurious (see Lemstrom 1904). Electrostimulation can also work in near darkness (Blackman et al. 1923) and must therefore be using stored food reserves. If the plant has little or no reserves, electrostimulation will not work and might even be injurious by using up reserves that it cannot afford. Electrostimulation might also divert resources from secondary to primary metabolism, thus weakening the plant's resistance to disease. It may be significant that in a large number of pot experiments that Blackman and Legg (1924) regarded as failures because electrostimulation reduced yield, the plants were heavily infected with mildew.

Finally, even when artificial electrostimulation does have an effect, if it is given when there is natural electrostimulation, the controls will grow faster so that the effects of treatment will seem to disappear. This would apply in conditions when the atmosphere becomes highly charged by thunderstorms or dust storms. Also, bearing in mind the sensitivity of seeds, if the seeds of the plants to be tested have been stored near electromagnetic fields (e.g., in an electrical refrigerator), they may have been stimulated already and no further effect may be possible.

Electrostimulation of Cultured Plant Tissues

The growth of tissue cultures is important in plant biotechnology, since, in many cases, genetically engineered cells first give rise to tissues; if the latter do not grow or fail to regenerate plantlets, they are of no value. Therefore, attempts have been made to improve plant cell growth in vitro by passing weak electrical currents through cultured tobacco callus (Rathore and Goldsworthy 1985a). The current was applied between a stainless-steel electrode in the callus and another in the culture medium, with the controls having similar electrodes, but left unconnected. There was up to 70% increase in growth rate with currents of 1 or 2 μA. The stimulation was dependent on the polarity of the applied current and only occurred when the callus was made negative to the culture medium.

Not all the electrical effects on tissue cultures are dependent on the polarity of the applied current. In similar experiments, but on a medium conducive to regeneration, Rathore and Goldsworthy (1985b) found up to sevenfold stimulation of shoot formation when currents of either polarity were used. In the original experiments, the stimulations were greater when the callus was negative to the medium, but this was later found to be due to the formation of toxic products in the callus from the positive electrode. Both polarities were equally effective when the current was delivered via salt bridges from Ag/AgCl electrodes (Rathore et al., 1988). This independence from the polarity of the applied current was reinforced by the later discovery that an alternating current was as effective as a direct current (Goldsworthy and Lagoa, unpublished).

The observation that an alternating current was equally effective stimulated attempts to make the technique less labor intensive. Instead of the laborious procedure of implanting electrodes individually into each callus, an alternating current was induced capacitatively in large numbers of calli at a time simply by putting the culture vessels between charged metal grids. Petri dish cultures were placed between two grids, with a 25 kV/m root mean square gradient at 50 Hz applied between them. Although it had no significant effect on the rate of shoot regeneration, the treated cultures had many more, but smaller, cells and also became much greener than controls placed between similar grids short-circuited together (Goldsworthy and Lagoa, unpublished). This suggests that the electrical field enhanced the rate of both cell and chloroplast division, but this effect may be separate from that on regeneration. The effect on greening is reminiscent of that observed in whole plants in electroculture (Blackman, 1924; Murr, 1963) and suggests that at least part of the mechanism of electrical stimulation of growth in cultured tissues may resemble that stimulation of growth in whole plants.

To have the maximum effect on organogenesis in tobacco tissue cultures, prolonged exposures to the electrical current were necessary and were routinely given for the entire growing period. Other workers, however, found stimulation of regeneration with relatively short exposures. For example, Dijak et al. (1986) reported that treating *Medicago* protoplasts for as little as 4 hours with 20 mV applied between silver electrodes 25 mm apart caused the formation of a large number of somatic embryos, whereas none appeared in the controls. Dramatic stimulation of growth and regeneration in colt cherry have been brought about by even shorter periods of exposure, such as the extremely brief high-voltage pulses used in electroporation (Ochatt et al. 1988). In these cases at least, the electrical treatment must be triggering some process leading to regeneration, which, once started, is self-sustaining. Perhaps the mechanisms involved may be the same as those for the electrostimulation of whole plants, which are also self-sustaining (see Blackman et al., 1923). They may, in fact, be a spin-off from natural selection for plants that respond to the weak electrical currents from thunderstorms by growth with increased vigor.

ELECTROSTIMULATION OF ANIMALS

There are no electrical stimulations of the growth of whole animals comparable with those of plants, perhaps because there is no selective advantage in their being able to respond in this way. There are, however, many effects of weak electrical currents on the regeneration of damaged tissues and organs, often associated with the polar growth and the migration of cells along the current path. In most, if not all, cases the effect seems to be due to the artificial current enhancing the effects of the weak natural currents that normally control regeneration.

Wound Healing

There are many reports of weak electrical currents assisting the healing of wounds in the skin of amphibians and mammals [see Vanable (1989)]. Normally, one electrode is placed in the wound and the other on a healthy part of the body. The flow of current appears to stimulate the migration of cells toward the wound and accelerates healing, apparently by enhancing the effect of natural wound currents.

Natural wound currents flow because the inside of the body is maintained at about 70 mV positive to the exterior by inwardly directed sodium pumps in the skin. When the skin is wounded, this "skin battery" is short circuited and current flows outward through the wound giving lateral voltage gradients of about 60–100 V/m in the skin immediately around the wound. This is sufficient to stimulate the migration of nearby cells toward the site of damage, since many of the cell types concerned have been shown to migrate in culture along gradients as low as 10 V/m. The polar migration of cells is not, however, the only controlling factor in wound healing, and there is also a chemotactic response of cells to chemicals produced in the inflamed area around the wound. Furthermore, there may be an electrical stimulation of healing that is independent of the polarity of the applied current, since accelerations of healing have been reported in mammals with currents of either polarity. Perhaps the relatively large currents flowing through a wound also cause a nonpolar increase in cellular activity to promote healing. If so, an artificial current that is in the wrong direction to cause the migration of cells toward the injury can still have a beneficial effect, but the cells would now have to rely on chemotaxis to send them in the right direction.

Regeneration of Limbs

The electrical stimulation of wound healing can sometimes extend to the regeneration of limbs. Borgens et al. (1977) amputated legs of adult frogs and applied -0.2 μA to the cut stump. They found that it stimulated the partial regeneration of a new leg, even though the adult frog is not normally able to replace lost limbs. This again may be because the applied current enhanced the current that normally flows out of wounds in the skin. This outward current makes the area immediately under the wound negative to the remainder of the body so that it behaves like a negative electrode. The effect is to attract new cells to the area. For example, isolated fibroblasts have been shown to migrate toward a negative electrode (Stump and Robinson 1982), and several workers have shown that the neurites of isolated nerve cells also grow in this direction (see

Nuccitelli 1984; Borgens 1989). It may be significant that the frogs' legs regenerating under the influence of an applied current described by Borgens et al. (1977) possessed an unusually large amount of nerve tissue.

The regeneration of limbs probably reflects a repetition of what happens during the development of an embryonic limb. This too is preceded by an outward flow of current in the regions concerned. It is unlikely, however, that the current determines the precise pattern of differentiation; it is too crude an instrument. The exact pattern is more likely to be defined by specific chemical determinants on the surface of each cell. For example, the pattern of innervation is probably controlled by different nerve cells following specific trails of chemical signals in the tissue. The main function of the current is probably to prevent the growing nerve cells accidentally doubling back and following the trail back to the starting point.

The electrical stimulation of limb regeneration may be true to a limited degree, even in humans. Weak electrical currents of the order of 10–20 $\mu A/cm^2$ have been measured, emerging from the stumps of amputated fingertips in children and, provided the current is not sealed off by stitching up the wound, new finger tips can be regenerated (Illingworth and Barker 1980).

Bone Healing

Another area where the application of weak electrical current has had beneficial effects is in the healing of bones. Bassett et al. (1964) reported a stimulation of new bone formation around the cathode when a pair of electrodes delivering currents of the order of microamps were implanted in dogs' femurs. Since then, many accelerations of fracture healing have been reported after placing electrodes delivering 10–20 μA in the bone on either side of the fracture [see McGinnis (1989) for a critical review]. Although there is a tendency for the formation of new bone to occur best on the cathodal side, this is not always the case. Sometimes the anode is best and alternating currents are also effective in promoting bone healing. Indeed some of the more modern noninvasive clinical techniques for treating non-union in fractures involve the induction of weak alternating currents in the bone by electromagnets placed outside the body. The stimulatory effect of alternating currents suggests that there may be another way in which the cells are activated electrically that is not directly concerned with their polar migration.

The effects of the applied current may be because it simulates natural potentials that occur after a fracture. Stern and Yageya (1980) placed electrodes directly on the bone and found a sharp peak in the bioelectrical potentials in the bone when it was fractured, but these fell to roughly their prefracture levels within a day. It seems possible that the initial fracture potential may help to initiate the bone repair process by activating the cells around the fracture site, but if, for some reason, this fails, it could result in non-union. If so, the artificial application of currents could supply an alternative stimulus to activate the healing process.

MECHANISMS OF ELECTROSTIMULATION

Importance of Calcium

There is almost certainly more than one mechanism by which artificially applied electrical currents stimulate cells, but calcium seems to be important in most if not all of them. It is well established that calcium ions are important regulators of metabolism in both animals and plants. They can activate ion channels responsible for much of the electrophysiology of cells, and they stimulate many "cascade" enzyme systems leading to increased metabolism and growth. Organisms can use calcium to regulate metabolism in many subtle and selective ways; for example, hormones often act by causing the release of calcium ions into the cytosol where they stimulate specific aspects of metabolism [see Hepler and Wayne (1985)]. In addition, the localized entry of calcium ions into the cell can cause polar growth, without which the evolution of large and complex multicellular organisms would not have been possible. Many of the effects of weak electrical currents on plants and animals appear to be due to their interference with calcium transport and consequent polar growth.

Electrical Control of Polar Growth

The electrical control of cell polarity has been well reviewed by Quatrano (1978) and Schnepf (1986). It has been studied mainly in the germinating zygotes of the seaweed *Fucus* and its close relative *Pelvetia*. The direction of zygote growth is normally determined by the direction of light. This causes an electrical current to flow into the shaded side, which precedes and predicts the direction of polar growth (Jaffe, 1966). Jaffe et al. (1974) showed the current to be carried by a number of different ions. These are pumped across the cell membrane by metabolically driven ion pumps at one end of the cell and leak back via passive ion channels at the other.

Several ion species contribute to the current, but possibly the most important in the primary establishment of polarity is calcium. Although calcium only carries a small proportion of the current, that proportion is largest at the onset of polarization (Jaffe et al., 1974), and its effects on metabolism are more dramatic than those of other ions (Hepler and Wayne, 1985). Further evidence is that when the calcium ionophore A23187 was supplied from one side of the zygote to allow localized calcium ingress, it initiated polar growth at the site of calcium entry (Robinson and Cone, 1980).

When the polarity of the zygote is being established, the current is at first very low, but it builds up over a period of hours with both more and different ions contributing, suggesting that calcium entry may be affecting the activity and location of channels for other ions. Initially, there may be several points at which the current enters on the shaded side of the cell, but these eventually form a single region that predicts the point of emergence of the rhizoid (Nuccitelli, 1978). The regions of current entry seem to correspond to regions of membrane containing ion channels, possibly linked by actin filaments (Brawley and Robinson, 1982, 1985). The mechanisms by which they are drawn together into a single region are unclear. It could, for example, be by the lateral electrophoresis of charged ion channels along the membrane in the transcellular current. Alternatively, it could be due to the mechanical activity of the actin filaments stimulated by the localized ingress of calcium. It is also conceivable that new channels are inserted into the membrane from within the cell at the site of current entry. Whatever the mechanism, the effect is that, in a matter of hours, a relatively large and stable current flows in at one end of the cell and out at the other.

Electrical Control of Physiological Polarity

The precise ways in which electrical polarity induces physiological polarity are also uncertain and likely to be complex. One mechanism is that the local ingress of calcium stimulates calcium-dependent metabolism, which, in turn, causes faster growth around its site of entry. Evidence for this is that relatively large amounts of free calcium are localized in the cytosol near the growing tips of *Fucus* rhizoid cells (Brownlee and Wood, 1986). Even minute amounts of free calcium ions inside the cell can cause major and prolonged stimulations of metabolism (Hepler and Wayne, 1985).

The localized stimulation of growth by calcium is unlikely to be the only effect, since it does not explain how the transcellular current might influence polar phenomena other than at the growing cell tip, nor does it explain the observation by Jaffe et al. (1974) that most of the current is carried by ions other than calcium. Another mechanism seems to be that the transcellular current brings about the electrophoresis of charged materials to specific regions of the cell and the nature of the ions carrying it is therefore not important. Electrophoresis could occur either in the cell interior or within the fluid mosaic of the membrane. Poo and Robinson (1977) provided evidence for the latter, when they showed that proteins moved along the external membrane of cells in weak artificially applied currents. In nature, proteins may be attracted along the membrane toward the oppositely charged poles of the cell, but because not all proteins have the same charge density, some will be attracted more powerfully than others. The highly charged ones may go all the way to the poles, and those with weaker charges will be displaced and forced to take up equilibrium positions some distance away. In this way, specific proteins can be located at defined positions along the length of its cell. Such a mechanism may determine the location of proteins initiating cell division near the cell's equator and may also enable daughter cells to have different cytoplasmic compositions when unequal cell division gives rise to differentiation.

Polarity in Multicellular Structures

The cells of multicellular structures also exhibit electrical polarity. Lund (1947a) showed that the filaments of the alga *Pithophora* were electrically polar, with the growing apex positive. This was because the apical end of each cell was positive to its base and together the cells contributed to the larger voltage gradient along the whole filament. The gradient along each cell implies that current must be flowing from its apex through the external medium to its base and completing the circuit via the interior. The neat nose-to-tail arrangement of the electrical polarities of the cells ensures that the filament elongates in a straight line and suggests that there must be some mechanism by which these polarities are coordinated.

Similar voltage gradients occur in higher plants. Wilkes and Lund (1947) found that the *Avena* coleoptile was electrically polar with the apex being negative to the base when measured by external electrodes. Electrodes placed in the lumen of the coleoptile, however, showed the opposite polarity. Perhaps, because the coleoptile is a terrestrial organ with no external medium through which a transcellular current can return, the current takes a circulatory path through the cells instead, with those on the inside being polarized in the opposite direction to those on the outside. The current, however, still passes longitudinally through the elongating region and could still define and coordinate the direction of cell elongation. Animals too have electrical patterns that control their development, but these tend to be much more complex than those of plants and involve control of the direction of cell migration as well as the control of cell growth (Nuccitelli, 1984).

Cell Polarity Can Change

Although the electrical polarities of most cells are relatively stable, probably due to the ion pumps and channels responsible becoming attached to the cytoskeleton (Quatrano, 1978), the currents can change direction when the direction of growth changes. For example, during the phototropic and gravitropic curvatures of cereal coleoptiles, a transverse component, which predicts the new direction of growth, develops in its normally longitudinal electrical field. The side toward which bending is to occur becomes negative (Schrank 1946, 1947a; Johnsson 1967), as if the electrical polarities of the component cells have become twisted with their positive basal ends now realigned partially toward the side. There is a corresponding change in the direction of auxin transport; instead of being toward the positive base of the coleoptile, it is now diverted toward its positive side. This region then grows faster so that the organ bends toward its negative side.

Artificial Currents Affect Cell Polarity

During the 1920s it was shown that an artificially applied current could polarize *Fucus* zygotes so that they germinated with their rhizoids pointing to the anode. Later work (Peng and Jaffe, 1976) showed that the direction of germination was not always consistent. Some batches of zygotes germinated toward the anode, some toward the cathode, whereas others germinated toward the anode in a high fields and toward the cathode in low fields. This resembles the pattern of electrically induced migration in some animal cells (Vanable, 1989), which is also toward the anode at high field strength and toward the cathode when the field strength is low.

The ability of cells to repolarize in an applied field may account for many of the effects of weak electrical currents on growth. For example, Schrank (1947b) showed that a transversely applied electrical current could realign the electrical polarities of cereal coleoptiles and this was associated with a change in the direction of growth. He applied currents of between 5 and 20 μA transversely for 2 min just below the apices of oat coleoptiles, then measured their own transverse potentials. For a few minutes after switching the current off, the coleoptiles retained the applied polarity (i.e., the area originally under the anode remained positive and that under the cathode remained negative), presumably due to the polarization of free ions in the coleoptile. This, however, was quickly replaced by a steady transverse potential, with the side that had been connected to the negative electrode now being positive. In effect, the coleoptile cells had repolarized so the direction of their own currents was the same as that of the current applied.

This effect has since been confirmed with the vibrating probe on cells from tobacco suspension cultures. When supplied for 2 min with an artificial current of a similar density to those used by Schrank, these cells also realigned their electrical polarities with most of their transcellular currents in the same direction as the current that had been applied (Mina and Goldsworthy, 1991).

Electrical Stimulation Changes the Direction of Auxin-Flow

The change in the direction of the natural current after the transverse application of an artificial current to the cereal coleoptile was followed by a bending response. This was apparently due to a redirection of auxin transport. Webster and Schrank (1953) showed that the electrically stimulated bending was dependent on indoleacetic acid (IAA) and did not occur in decapitated coleoptiles from which the natural IAA supply had been removed. It was, however, restored if the tip was replaced by an agar block containing IAA, even if this was delayed for more than 2 hr after switching off the current. These findings indicate that the transverse current realigned the electrical polarities of the coleoptile cells with their positive poles redirected toward the cathode. Auxin (which is normally transported toward the positive base of the cells) was now transported toward its new positive poles facing the cathode. Consequently, the cathodal side of the coleoptile accumulated more auxin and grew faster causing a curvature toward the anode.

Electrically Induced Polar Auxin-Flow Stimulates Growth in Cultured Tissues

A similar mechanism to that in the electrically treated coleoptile may have caused the electrostimulation of tobacco callus growth reported by Rathore and Goldsworthy (1985a). This also involved an apparent polar transport of auxin toward the cathode (Goldsworthy and Rathore, 1985), evidence being that the stimulation only occurred when the callus was made negative to the culture medium, was dependent on a supply of IAA in that medium, and was inhibited by tri-iodobenzoic acid (an inhibitor of the polar transport of auxin). Furthermore, it did not occur if the IAA in the culture medium was replaced by 2,4-dichlorophenoxyacetic acid (an auxin that normally shows little or no polar transport). Subsequent work (Goldsworthy and Lagoa, unpublished) showed that the natural electrical polarity of an untreated callus is apex negative to base, so that normal polar transport would tend to exclude auxin from the culture. Applying a negative electrode at the apex would therefore reverse this polarity, promote the polar transport of auxin into the tissue, and produce the observed stimulation of growth. Applying a current in the opposite direction simply reinforced the natural polarity and so had relatively little effect.

Cells May Control Each Other's Polarities Electrically

The ability of plant cells to reorient their polarities in line with an externally applied current may be due to its effects on a natural mechanism, which enables cells to coordinate their polarities so that the positive pole of one cell aligns itself adjacent to the negative pole of its neighbor. This would give a tissue in which the cells shared the same polarity and could therefore grow in an organized fashion. There is evidence that animal cells may control the polarity of their neighbors in much the same way. Lund (1947b) described an experiment in which the coelenterate *Obelia* was mechanically disrupted and its cells passed through bolting cloth. Normally, such cells reassociate to form new animals, implying that they have some means of coordinating one another's polarities. When the reassociation was performed in a weak electrical current, however, the new organisms regenerated with the polyps pointing toward the anode. This suggests that the applied current coordinated the polarity of the cells as they rejoined and that the natural mechanism coordinating the reassociation may be the cells' own polar currents.

Does Lateral Electrophoresis Realign Cell Polarities?

One explanation for the realignment of cell polarities in external electrical fields may lie in the electrophoretic mobility of ion pumps and channels in the cell membrane. For example, if the ion pumps or channels that make the basal end of a cell positive are themselves positively charged, they would be attracted toward the negatively charged apex of the cell beneath. This does not, however, explain the observation that some cells can be polarized by alternating currents. Novak and Bentrup (1973) showed that alternating currents could initiate polar growth in the *Fucus* zygote, with the rhizoids pointing at either electrode in equal proportions. Since the length of each half-cycle was insufficient to allow a significant electrophoretic movement of cell components before its effect was reversed by the next half-cycle, they concluded that the initial polarizing effect was not due to electrophoresis.

Do Differences in Membrane Potential Align Cell Polarities?

Novak and Bentrup (1973) suggested that electrically induced polarization in the *Fucus* zygote was due to the cell sensing differences between the membrane potential at either end. Because the interior of the cell has a low resistance and the cell membrane a very high one, the bulk of the voltage applied across a cell appears across its membranes. This adds to the membrane potential at one end of the cell and subtracts at the other. They calculated that artificial currents strong enough to cause polarization changed the membrane potential at either end of the cell by about 4.5 mV (about 10%). The main problem with this hypothesis is that the difference due to the applied voltage is still small compared with the membrane potential and much smaller than the natural changes in membrane potential that occur as germination proceeds. For this mechanism to be effective, the cell must have the means to compare its membrane potentials in all regions at the same instant and use the result to control the activity or location of the ion pumps and channels responsible for inducing electrical polarity.

This prompts the questions, How is this done? and would such a mechanism be sensitive enough to account not only for the polarization of the *Fucus* zygote, but also for the more sensitive responses of other cells to artificially applied fields? These would include the electrically induced migration of animal cells, where the voltage across the cell may be only about 1 mV (Cooper and Schliwa, 1986) and the stimulation of embryo formation from plant protoplasts reported by Dijak et al. (1986) with a potential difference of only 20 mV between electrodes 25 mm apart. The answer is that such sensitivities to applied electrical fields are possible. The means to detect extremely weak electrical currents occur and have been studied in the electroreceptor cells of a wide variety of animals. They appear to have evolved independently several times and may simply be more highly developed versions of a mechanism that evolved in all cells that use electrical currents to control their growth and migration.

Electroreceptors

Extremely sensitive electrical sensors are common in animals. Specialized electroreceptors are found in fish, certain amphibia, and even primitive mammals. Many fish can navigate by sensing the direction of the tiny currents that flow through their bodies as they swim through the Earth's magnetic field. Others use them to detect prey. Among the most sensitive are those of the shark family that can find their prey by sensing the electrical activity of its nerves and muscles. The best of them can detect

AC and DC voltage gradients in the surrounding water of about 0.5 μV/m (about a thousand times more sensitive than a vibrating probe). They are normally based on groups of special receptor cells concealed in cavities in the skin that can sense changes in the potential difference across their membranes of as little as 0.2 μV. The simplest are probably the "tonic" receptors of fresh water teleosts, whose receptor cells have voltage-gated channels in their membranes that respond to a slight depolarization on one face of the cell by allowing calcium ions to enter. The entry of the calcium does not depolarize the membrane and causes an action potential because calcium-gated potassium channels allow an outward flow of potassium to counterbalance the effect. The result is that the membrane potential remains stable, but the entry of calcium triggers the production of a neurotransmitter that stimulates the generation of action potentials in neighboring nervous tissue (Bennet and Obara, 1986).

Calcium Channels as Electroreceptors in Ordinary Cells

If the relatively sophisticated electroreceptor mechanisms of fish and other animals evolved from those coordinating the polarities and stimulating polar growth and migration of ordinary cells, it is possible that such ordinary cells also respond to electrical stimulation by allowing calcium ingress. The mechanism by which they respond to minute differences in membrane potential at either end of the cell is far from clear, but a simple model consisting of voltage-gated calcium channels and calcium-gated channels for potassium interspersed with each other over the cell surface is one possibility. If an electrical current is applied across such a cell, the region of membrane nearest the cathode would tend to depolarize. The voltage-gated calcium channels in this region would open more frequently to let in calcium, but this would result in the calcium-gated potassium channels opening to allow potassium efflux and restore the membrane potential. The effect would be that moderately large quantities of calcium could enter without depolarizing the membrane and causing an action potential. Once inside, because of the ease with which it binds to proteins, free calcium would be restricted to the area immediately around its site of entry where it would stimulate metabolism. Calcium ions are renowned for their ability to induce "cascade-type" reactions in which small amounts of calcium rapidly trigger massive metabolic events (Hepler and Wayne, 1985). Because of this amplifying effect, small differences in calcium concentration at either end of the cell could cause relatively large differences in other metabolites and so induce cell polarity.

Calcium Ingress May Affect Actin

One effect of calcium ingress may be to stimulate the localized activity of actin molecules, as proposed by Cooper and Schliwa (1986), to explain a calcium-dependent migration of fish keratocytes toward the cathode in an applied field. Actin might also be involved in the active translocation of raw materials to support localized polar growth within the cell and also in the translocation and binding of ion channels to the cytoskeleton at the site of calcium ingress to stabilize cell polarity once it has been induced (Brawley and Robinson, 1982, 1985).

Cells with Anodal Responses

The examples already discussed are of cells that migrate or grow toward the cathode. Other cells grow or migrate only to the anode, whereas still others may respond in either direction. For example, some batches of *Fucus* zygotes grow toward the cathode at low field strength and to the anode at high field strength (Peng and Jaffe, 1976). A similar situation has been found with the migration of animal cells (Vanable, 1989). The observation that at higher field strengths some cells grow or migrate toward the anode requires a slight modification to the hypothesis presented earlier. All that needs to be postulated, however, is that there is a large number of voltage-gated calcium channels that are normally closed but that open when the membrane becomes strongly hyperpolarized. In an electrical field, the side adjacent to the anode becomes hyperpolarized. If a strong field opened these channels and they were either more active or in greater number than those that respond to depolarization, the anodal side of the cell would become the main site of calcium entry and the cell would now grow (or migrate) toward the anode. The differing directional responses of different batches of *Fucus* zygotes to applied fields reported by Peng and Jaffe (1976) could then be ascribed to their having differing proportions of the two types of calcium channel.

There are considerable advantages to a cell being able to control its direction of growth in an applied field. For example, a nerve cell must show a cathodal response if it is to grow toward the periphery of the organism to innervate a limb. Similarly, other cells migrating to the site of limb formation or to repair a wound in the skin may also show a cathodal response because the wound is negative to the rest of the body. When cells unite to form a tissue, however, an anodal response may be required. Cells with an anodal response will have their main site of current entry facing the anode; that is, the negative end of the cell faces the anode. This is what is required if cells are to unite to form a tissue in which all the cells share the same polarity, since the positive end of one cell must locate with the negative end of its neighbor.

Nonpolar Responses to Electric Currents

Not all the effects of electrical currents on cell growth and differentiation can be explained easily in terms of changes in cell polarity, especially when alternating currents are effective or when the response is not always related to the direction of the applied current; for example, the electrostimulation of the growth of whole plants (Blackman and Legg, 1924), the regeneration of shoots in cultured plant tissues (Rathore and Goldsworthy, 1985b), and the regeneration of animal skin and bone (Borgens et al. 1989). Do these involve a fundamentally different mechanism? Perhaps not. If we accept that electrically stimulated calcium entry is responsible for initiating and maintaining polar growth, there is no reason why it should not also activate nonpolar responses. Although it must be accepted that, because of their ability to bind non-specifically to other molecules, free calcium ions are likely to be restricted to the region around their site of entry, calcium in the form of its active calmodulin complex is much more mobile. Once formed, it could be active almost everywhere in the cell, as indeed it would probably have to be if increased metabolism in other parts of the cell is required to provide the raw materials for polar growth. It is therefore conceivable that many of the effects of weak electrical currents on cells may be due to a stimulation by calcium of their normal function unrelated to polar growth.

Evolution of the Electrophysiology of Growth

Possibly, the varied effects of electrical currents on cells and organisms can be summarized by trying to trace the evolution of the mechanisms that gave rise to them. The link with calcium must have begun very early in evolution. All known living organisms depend on organic phosphates as metabolic intermediates, but many of these are precipitated in the presence of calcium ions. Early cells, like those of today, had to pump out calcium to stay alive. Calcium ingress was a sign of life-threatening damage to the external membrane, and natural selection produced cells that responded with an immediate stimulation of metabolism to provide materials and energy for repair.

Controlled calcium entry then became the simplest and most convenient basis for regulating growth and metabolism for other purposes. Polar growth and cell migration was perhaps the first of these (Goldsworthy, 1988). Consider a primitive cell growing on the surface of a nutrient substrate. Natural selection would favor any mechanism that allowed that cell to either grow into or move into the substrate, but it needed a signal to tell it the right direction. An electrical signal was already available because the cell would probably be absorbing nutrients by ion cotransport mechanisms, which tend to depolarize the membrane adjacent to the substrate. All that had to evolve was a voltage-gated ion channel that would allow the entry of calcium to stimulate metabolism leading to growth specifically in this part of the cell. In this way, simple spherical cells could evolve into food-seeking filamentous structures like the mycelia of fungi. Other organisms came to respond by moving in the direction of the calcium stimulus, and these became the forerunners of the animal kingdom. The ability of cells to orient their growth and locomotion in weak electrical currents was also a major step in the development of multicellular organisms in which the direction of growth of the individual cells is controlled and coordinated to form organized structures by the currents generated by others.

A problem that arises with calcium-induced polar growth or locomotion is that it requires an infrastructure of metabolism in unstimulated parts of the cell to supply the raw materials and energy. This cannot be achieved by the diffusion of free calcium ions from its site of entry because they are too readily bound to other molecules. The calmodulin molecule, which carries calcium in a metabolically active form, may have evolved to transmit this signal from the site of entry to other regions very early in evolution. Consequently, localized calcium entry in any part of the cell might be expected to stimulate metabolism, even in cells not necessarily capable of polar growth. Natural selection for responses to external electrical fields not directly related to polar growth or migration along the line of the current may be responsible for the increases in metabolism and growth in seeds and plants exposed to large atmospheric voltage gradients and also for the extreme electrosensitivity of many fish.

All this has consequences for our ability to modify the behavior of organisms and cells by the artificial application of weak electrical currents. These are likely to cause calcium ions to enter cells and result in long-lasting stimulations of metabolism as well as to initiate polar growth or locomotion. It can be said in conclusion that in some cases, the application of weak artificial currents can assist development, regeneration, and healing in both animal and plant tissues because they enhance the effects of natural currents promoting growth. In some cases, this may be further improved by adjusting the direction of the current so that it directs polar cell growth and migration along its line of application.

SUMMARY

Significant stimulations of plant growth can occur following exposure to strong electric fields. Attempts were made in the early part of the century to exploit this by the "electroculture" of crops beneath wires carrying high voltages. The results were not always consistent and success appeared to depend on the weather. A modern ecological explanation is that plants use the dramatic increases in atmospheric electric fields which often precede rainfall to stimulate growth and make best use of an anticipated supply of water. The failure of some of the electroculture experiments may be due to the indiscriminate use of the fields and the induction of rapid growth under stressful environmental conditions when this would lead to injury. More recently, electrical stimulations of growth and development have been confirmed in plant tissue cultures and even isolated protoplasts. This indicates that the electrical effect on growth is at the level of individual cells and does not require the intact organism.

There are no reports of stimulations of whole-animal growth by artificially-applied electric currents, but there are many examples of stimulations of the repair of injuries. These include the healing of wounds, the union of bone fractures and even the regeneration of lost limbs and appendages. Present evidence suggests that artificially-applied electric currents often affect cell growth and their metabolic functions by interacting with the mechanisms by which natural currents control these processes. Natural currents, driven through cells by ion pumps in their membranes, control the direction of cell growth in animals and plants and (in animals) cell migration. For example, metabolically driven ion pumps in the skin of animals make the inside of the body electrically positive. If the skin is wounded, positive current leaks out, the flow of current towards the injury directs the migration of surrounding cells to repair the damage and electrodes placed on the wound can assist the process.

There is also a relationship between the flow of electric currents through tissues and the control of cell growth by calcium. It is believed that eukaryotic cells sense electric currents by a mechanism which involves the entry of calcium ions from the surrounding medium. In addition to telling them the direction of the current so that they can coordinate their growth, the additional calcium stimulates metabolism to provide the extra energy and materials for the response. Artificially applied currents also do this and it provides a further mechanism by which healing in animals is enhanced by artificially applied currents and the yield of crops can be increased by strong electric fields.

References

Bassett, C. A. L., Pawluk, R. J., and Becker, R. O. (1964). Effects of electric currents on bone. Nature 204: 652–654.

Bennet, M. V. L., and Obara, S. (1986). Ionic mechanisms and pharmacology of electroreceptors. Pages 157–181. In Electroreception. Bullock, T. H., and Heiligenberg, W., eds. John Wiley, New York.

Blackman, V. H. (1924). Field experiments in electroculture. J. Agricult. and Sci. 14: 240–267.

Blackman, V. H., and Legg, A. T. (1924). Pot culture experiments with an electric discharge. J. Agricult. and Sci. 14: 268–273.

Blackman, V. H., Legg, A. T., and Gregory, F. G. (1923). The effect of direct electric current of very low intensity on the rate of growth of the coleoptile of barley. Proc. Roy. Soc. Lond. B. 95: 214–228.

Borgens, R. B. (1989). Artificially controlling axonal regeneration and development by applied electric fields. Pages 117–170. In Electric Fields in Vertebrate Repair. Borgens, R. B., Robinson, K. R., Vanable, J. W., and McGinnis, M. E., eds. Alan R. Liss, New York.

Borgens, R. B., Robinson, K. R., Vanable, J. W., and McGinnis, M. E. (1989). Electric Fields in Vertebrate Repair. Alan R. Liss, New York.

Borgens, R. B., Vanable, J. W., and Jaffe, L. F. (1977). Initiation of frog limb regeneration by minute currents. J. Exptl. Zool. 200: 403–416.

Brawley, S. H., and Robinson, K. R. (1982). Rhizoid formation in fucoid embryos is accompanied by F-actin formation. J. Cell Biol. 95: Abstract page 152a.

Brawley, S. H., and Robinson, K. R. (1985). Cytochalasin treatment disrupts endogenous currents associated with cell polarization in fucoid zygotes: Studies on the role of F-actin in embryogenesis. J. Cell Biol. 100: 1173–1184.

Briggs, L. J., Campbell, A. B., Heald, R. H., and Flint, L. H. (1926). Electroculture, U.S. Dept. Agriculture, Bulletin 1379.

Brownlee, C., and Wood, J. W. (1986). A gradient of cytoplasmic free calcium in growing rhizoid cells of Fucus serratus. Nature 320: 624–626.

Chalmers, J. A. (1957). Atmospheric Electricity. Pergamon Press, London.

Cooper, M. S., and Schliwa, M. (1986). Transmembrane Ca^{2+} fluxes in the forward and reversed galvanotaxis of fish epidermal cells. Pages 311–318. In Ionic Currents in Development. Nuccitelli, R., ed. Alan R. Liss, New York.

Dijak, M., Smith, D. L., Wilson, T. J., and Brown, D. C. W. (1986). Stimulation of direct embryogenesis from mesophyll protoplasts of Medicago sativa. Plant Cell Rep. 5: 468–470.

Ellis, H. W. (1981). The effect of electricity on Hordeum vulgare and Escherichia coli. Ph.D. Thesis, University of London.

Ellis, H. W., and Turner, E. R. (1978). The effect of electricity on plant growth. Sci. Prog. Oxf. 65: 395–407.

Goldsworthy, A. (1988). Growth control in plant tissue cultures. Pages 35–52. In Biotechnology in Agriculture. Mizrahi, A., ed. Alan R. Liss, New York.

Goldsworthy, A., and Rathore, K. S. (1985). The electrical control of growth in plant tissue cultures: The polar transport of auxin. J. Exptl. Bot. 36: 1134–1141.

Hepler, P. K., and Wayne, R. (1985). Calcium and plant cell development. Ann. Rev. Plant Physiol. 38: 397–439.

Illingworth, C. M., and Barker, A. T. (1980). Measurement of electrical currents emerging during the regeneration of amputated finger tips in children. Clin. Phys. Physiol. Meds. 1: 87–89.

Jaffe, L. F. (1966). Electrical currents through the developing Fucus egg. Proc. Natl. Acad. Sci. USA 56: 1102–1109.

Jaffe, L. F., Robinson, K. R., and Nuccitelli, R. (1974). Local cation entry and self-electrophoresis as an intracellular localization mechanism. Ann. N.Y. Acad. Sci. 238: 372–389.

Johnsson, A. (1967). Relationships between photo-induced and gravity-induced electrical potentials in Zea mays. Physiol. Plant. 20: 562–579.

Lemstrom, S. (1904). Electricity in Agriculture and Horticulture. The Electrician. Printing and Publishing Co., London.

Lund, E. J. (1947a). Polar distribution of maintained electric circuits on the surface of Pithophora sp. cells. Pages 1–15. In Bioelectric Fields and Growth. Lund, E. J., ed. University of Texas Press, Austin, TX.

Lund, E. J. (1947b). Control of orientation of growth in reassociating cells of Obelia. Pages 231–233. In Bioelectric Fields and Growth. Lund, E. J., ed. University of Texas Press, Austin, TX.

Lutkova, I. N., and Oleshko, P. M. (1965). Effect of electric current during stratification in cherry seeds. (in Russian) Fiziol. Rast. 12: 238–241.

Malan, D. J. (1963). Physics of Lightning. The English Universities Press, London.

McGinnis, M. E. (1989). The nature and effects of electricity in bone. Pages 225–284. In Electric Fields in Vertebrate Repair. Borgens, R. B., Robinson, K. R., Vanable, J. W., and McGinnis, M. E., eds. Alan R. Liss, New York.

Mina, M. G., and Goldsworthy, A. (1991). Changes in the electrical polarity of tobacco cells following the application of weak external currents. Planta 186: 104–108.

Murr, L. E. (1963). Plant growth response in a stimulated electric field environment. Nature 200: 490–491.

Novak, B., and Bentrup, F. W. (1973). Orientation of fucus egg polarity in electric A.C. and D.C. fields. Biophysics 9: 253–260.

Nuccitelli, R. (1978). Ooplasmic segregation and secretion in the Pelvetia egg is accompanied by a membrane-generated electrical current. Dev. Biol. 62: 13–33.

Nuccitelli, R. (1984). The involvement of transcellular ion currents and electric fields in pattern formation. Pages 23–46. In Pattern Formation: A Primer in Developmental Biology. Malacinski, G. M., and Bryant, S. V., eds. Macmillan, New York.

Ochatt, S. J., Chand, P. K., Rech, E. L., Davey, M. R., and Power, J. B. (1988). Electroporation-mediated improvement of plant regeneration from colt cherry (Prunus avium X pseudocerasus) protoplasts. Plant Sci. 54: 165–169.

Peng, H. B., and Jaffe, L. F. (1976). Polarization of fucoid eggs by steady electrical fields. Dev. Biol. 53: 277–284.

Pittman, U. J., and Ormrod, D. P. (1970). Physiological and chemical features of magnetically treated winter wheat seeds and resultant seedlings. Can. J. Plant Sci. 50: 211–217.

Poo, M. M., and Robinson, K. R. (1977). Electrophoresis of concanavalin A receptors along muscle cell membranes. Nature 265: 602–605.

Quatrano, R. S. (1978). Development of cell polarity. Ann. Rev. Plant Physiol. **29**: 487–510.

Rathore, K. S., and Goldsworthy, A. (1985a). Electrical control of growth in plant tissue cultures. Bio/Technol. **3**: 253–254.

Rathore, K. S., and Goldsworthy, A. (1985b). Electrical control of shoot regeneration in plant tissue cultures. Bio/Technol. **3**: 1107–1109.

Rathore, K. S., Hodges, T. K., and Robinson, K. R. (1988). A refined technique to apply electrical currents to callus cultures. Plant Physiol. **88**: 515–517.

Robinson, K. R., and Cone, R. (1980). Polarization of fucoid eggs by a calcium ionophore gradient. Science **207**: 77–78.

Schnepf, E. (1986). Cellular polarity. Ann. Rev. Plant Physiol. **37**: 23–47.

Schonland, B. F. J. (1928). The interchange of electricity between thunderclouds and the earth. Proc. Roy. Soc. **A118**: 252–262.

Schrank, A. R. (1946). Note on the effect of unilateral illumination on the transverse electrical polarity of the *Avena* coleoptile. Plant Physiol. **21**: 362–365.

Schrank, A. R. (1947a). Analysis of the effects of gravity on the electric correlation field in the coleoptile of *Avena sativa*. Pages 75–123. In Bioelectric Fields and Growth. Lund, E. J., ed. University of Texas Press, Austin, TX.

Schrank, A. R. (1947b). Electrical and curvature responses of the *Avena* coleoptile to transversely applied direct current. Pages 217–231. In Bioelectric Fields and Growth. Lund, E. J., ed. University of Texas Press, Austin.

Shatilov, F. V., and Trifonova, M. F. (1968). The effect of a direct current on the metabolism of sprouting barley seeds. (in Russian). Electron Orab. Mater. **1**: 67–74.

Stanko, S. A., and Koshevnikova, N. F. (1972). Physiologico-biochemical changes in wheat plants before and after pre-sowing treatment of seeds with an electric current. (in Russian). Sel'Sko-Khozyaistvennaya **7**: 624–626.

Stern, L. L., and Yageya, J. (1980). Bioelectric potentials after fracture of the tibia in rats. Acta Orthop. Scand. **51**: 601–608.

Stump, R. F., and Robinson, K. R. (1982). Directional movement of *Xenopus* embryonic cells in an electric field. J. Cell Biol. **95**: Abstract page 331a.

Vanable, J. W. (1989). Integumentary potentials and wound healing. Pages 171–224. In Electric Fields in Vertebrate Repair. Borgens, R. B., Robinson, K. R., Vanable, J. W., and McGinnis, M. E., eds. Alan R. Liss, New York.

Webster, W. W., and Schrank, A. R. (1953). Electrical induction of lateral transport of 3-indoleacetic acid in the *Avena* coleoptile. Arch. Biochem. Biophys. **47**: 107–118.

Wilkes, S. S., and Lund, E. J. (1947). The electric correlation field and its variations in the coleoptile of *Avena sativa*. Pages 24–75. In Bioelectric Fields and Growth. Lund, E. J., ed. University of Texas Press, Austin, TX.

CHAPTER 13

Stimulation of Plant Cell Division and Organogenesis by Short-Term, High-Voltage Electrical Pulses

M. R. Davey

N. W. Blackhall

K. C. Lowe

J. B. Power

ABSTRACT

High-voltage, short-duration electrical pulses stimulate DNA synthesis in isolated higher plant protoplasts. They also promote the growth of protoplast-derived cells, and shoot regeneration from protoplast-derived tissues. Such effects of electrostimulation persist over many cell generations. This enhancement of growth and organogenesis has application in the multiplication of elite individuals and in maximizing the recovery of genetically engineered plants following somatic hybridization and transformation. Detailed knowledge of the precise mechanisms of action of electrical pulses on the stimulation of growth and morphogenesis in plant cells are still lacking. The possible synergistic effects of electrical and chemical parameters require further investigation.

INTRODUCTION

A unique feature of plant cells and tissues in culture is that under the correct chemical and physical conditions, such cells express their totipotency. Thus, a single cell is capable of undergoing mitotic division to produce a tissue from which one or more shoots can be regenerated through organogenesis and/or somatic embryogenesis. These regenerated shoots are capable of being grown to mature, flowering plants, which, if fertile, will produce viable seed.

The totipotency of plant cells enables them to be used as source material for the genetic manipulation of plants through the introgression of genes by somatic hybridization, cybridization, and transformation. For example, isolated plant protoplasts (naked cells) can be fused chemically and electrically to produce somatic hybrids to circumvent sexual incompatibility at the intergeneric and interspecific levels. Similarly, protoplast fusion can be used to transfer organelle-specific genomes from donor to recipient protoplasts for cybrid (cytoplasmic hybrid) production (Davey and Kumar, 1983). The introduction of specific DNA sequences into plants through transformation is possible by direct DNA uptake into isolated protoplasts, by *Agrobacterium*-mediated gene delivery into cells and tissues, and by particle bombardment of target tissues. Since the success of all of these approaches necessitates maximum shoot regeneration from genetically manipulated material, any chemical or physical parameters that facilitate and stimulate plant regeneration are of significance.

During the mid 1980s, weak electrical currents of prolonged duration over several days were shown to stimulate growth and shoot regeneration from cultured tobacco callus (Rathore and Goldsworthy, 1985a), through promotion of the polar transport of auxin in the tissue (Rathore and Goldsworthy, 1985b). A more extensive review of the effects of such electrical currents of long-term duration on plant cells is presented by Goldsworthy in Chapter 12 of this volume. Concurrently, high-voltage, short-duration electrical pulses were also being used to increase the permeability of the plasma membrane to molecules of various sizes, including DNA, in both cultured animal cells (Neumann et al., 1982; Reiss et al., 1986; Toneguzzo and Keating, 1986) and isolated plant protoplasts (Hashimoto et al., 1985; Shillito et al., 1985; Fromm et al., 1986) through electrically induced pore formation (electroporation). Circumstantial evidence also indicated that some isolated plant protoplasts divided more rapidly in culture following electroporation than untreated protoplasts.

Such observations provided the incentive to investigate, in greater detail, the effects of high-voltage electrical pulses on the growth in culture of plant cells. A range of commercially available and purpose-built instruments has been reported for generating electrical pulses. The merits and limitations of this equipment are discussed by Jones et al. in Chapter 1 of this volume.

STIMULATION OF DIVISION OF PROTOPLAST-DERIVED CELLS

Rech et al. (1987) reported the results of extensive investigations of the effect of electrical field pulses, ranging from 250 to 2000 V at 10–50 nF capacitance and 10 to 50 μsec duration, on the growth in culture of isolated protoplasts of the wild soybean, *Glycine canescens*, and the woody species *Prunus avium* x *pseudocerasus*, *Pyrus communis*, *Solanum dulcamara*, and *S. viarum*. Protoplasts were isolated enzymatically from seedling cotyledons (*Glycine*, *S. viarum*), cell suspensions (*P. avium* x *pseudocerasus*) and callus (*P. communis*) before being suspended at $1.0 \times 10^5 - 2.0 \times 10^6$ mL^{-1} in 5.0-mM 2-N-morpholinoethane sulfonic acid (MES)-based buffer solution containing 6-mM magnesium chloride and 0.5–0.7-M mannitol at pH 5.8. Protoplasts were pulsed in the chamber of a commercially available electroporator (Dia-Log, Dusseldorf 13, FRG) and exposed to three successive exponential pulses separated by 10-sec intervals. Following electroporation, protoplasts were diluted with, and maintained in, their appropriate culture media. In all cases, the percentage of viable protoplasts decreased with increasing voltage and capacitance, lower voltages of long pulse length (e.g., 250 V, 50 nF) being more detrimental to protoplast viability than higher voltages of short duration (e.g., 1250 V, 10 nF). The size of protoplasts also influenced their ability to survive electrical pulses; protoplasts smaller than 20 μm in diameter, such as those of *Prunus*, being less sensitive to electrical pulses than protoplasts of twice this diameter (e.g., those of *Pyrus*). *Glycine*, *S. dulcamara*, and *S. viarum* protoplasts of mean diameters 23, 28, and 37 μm, respectively, exhibited an intermediate response. Protoplasts electroporated with pulses up to 1250 V regenerated new cell walls and also entered mitotic division within 5 days of culture, this earlier onset of division being most noticeable in protoplasts of *Prunus*, *Pyrus*, and *S. viarum*, which, in the absence of electrical treatment, had long lag phases of 15, 9, and 11 days, respectively, before mitosis. In general, pulses of 1500 V or greater inhibited viability and growth of the protoplast systems assessed.

   In addition to stimulating the division of protoplast-derived cells, electroporation increased significantly the plating efficiency, that is, the ability of protoplasts to form cell colonies. For example, in comparison to untreated protoplasts of *Prunus* that had only undergone their first mitotic division when assessed after 10 days of culture, protoplasts electroporated at voltages up to 1750 V for short pulses (10 and 20 nF) had already developed into cell colonies when assessed at this time. The plating efficiency was inversely related to the pulse duration, long pulses from a 50-nF capacitor giving less stimulation than shorter pulses at the same voltage. The number of protoplast-derived cells per colony, indicative of the rate of cell division, was greatest at lower voltages and longer pulses; protoplasts electroporated at 250 V and 40 nF had divided 5 times within 10 days of culture, whereas untreated protoplasts exhibited sporadic first division. This enhancement of cell division was sustained throughout culture, with electroporated protoplasts giving significantly more colonies compared to untreated protoplasts of the same species.

The ability to electrostimulate protoplasts of *Glycine*, *Prunus*, *Pyrus*, and *Solanum*, particularly those of the woody genera *Prunus* and *Pyrus*, which had earlier proven to be recalcitrant in culture, encouraged workers to apply electroporation to protoplasts of cereals, which, to that time and, indeed, to date, generally remain difficult to culture. Working with protoplasts isolated from embryogenic cell suspensions of *Pennisetum squamulatum* (pearl millet), Gupta et al. (1988) subjected protoplasts to 100, 250, and 500 V at 40 nF with six pulses at 10-sec intervals in a buffer similar to that used by Rech et al. (1987) and the same electroporator as used by the latter workers. As in the case of *Glycine*, *Prunus*, *Pyrus*, and *Solanum*, electroporated protoplasts of *Pennisetum* divided within 4 days of culture compared with untreated controls, which required 5–7 days before entering division. The number of electroporated protoplasts dividing at day 10 was twice that of untreated protoplasts, with protoplasts plated at 2.0 and 3.0×10^5 mL^{-1} giving a two- to threefold increase in the number of protoplast-derived colonies compared to untreated controls. Electroporation at 250 V and 40 nF with six pulses was optimal for protoplast division and colony formation.

Other workers have investigated the effects of a range of electrical parameters on isolated protoplasts and found electroporation to enhance protoplast division. Montane and Teissé (1992) studied the influence of field strength, pulse duration, pulse number, and protoplast size on the division of isolated tobacco protoplasts. These workers confirmed the observations of Rech et al. (1987) that electrical pulses increased the plating efficiency of protoplasts up to 50% above that of untreated controls. In the case of tobacco, there was a linear increase in plating efficiency up to 0.8 KV cm^{-1}. Both the osmotic pressure of the electroporation buffer and protoplast size were found to be important chemical and physical parameters, larger tobacco protoplasts being lysed by the electrical pulses before the smaller ones were stimulated. Montane and Teissé (1992) emphasised that, to achieve the maximum benefit from electrical treatment, it was necessary to fractionate heterogeneous protoplast preparations into homogeneous subpopulations, a subpopulation with a mean diameter of 25 ± 5 μm being produced using a Ficoll 400 gradient in their experiments. Removal of the larger protoplasts also reduced the amount of cellular debris in the preparations caused by lysis of the larger protoplasts during electroporation. This reduced the release of toxic compounds, which can inhibit division of surviving protoplasts.

ELECTROPORATION STIMULATES DNA SYNTHESIS IN PLANT PROTOPLASTS

The observation that pulsing electromagnetic fields stimulated DNA synthesis in cultured mammalian cells (Rodan et al., 1978; Liboff et al., 1984; Takahashi et al., 1986) prompted Rech et al. (1988) to investigate whether electroporation could stimulate DNA synthesis in plant protoplasts. Protoplasts of *Prunus avium* x *pseudocerasus* and *Solanum dulcamara*, which divided after 4 and 3 days when electroporated compared to 15 and 6 days, respectively, when not treated in this way, were pulse labeled by incubation for 1.5 hr with [^{3}H]thymidine and samples taken at 24, 48, 72, and 96 hr after electroporation. The maximum incorporation of [^{3}H]thymidine into acid precipitable material occurred after 24 hr for *Prunus* and at 24 and 72 hr for *Solanum*, these two peaks of incorporation suggesting an oscillation in timing of DNA synthesis in the latter case. Protoplasts of both genera not subjected to electroporation showed a significantly lower incorporation of radiolabeled thymidine over the same period.

In continuous labeling experiments, where both electroporated and untreated protoplasts were incubated with [^{3}H]thymidine for up to 48 hr, the electroporated protoplasts exhibited increased incorporation of thymidine compared to controls. Such results confirmed that electroporation promoted DNA synthesis in higher plant protoplasts, which, in turn, was probably associated with the earlier onset of mitosis in protoplast-derived cells.

ELECTROPORATION OF ISOLATED PROTOPLASTS STIMULATES PLANT REGENERATION FROM PROTOPLAST-DERIVED TISSUES

A noticeable feature of culture vessels containing electroporated protoplasts was the increased size and number of protoplast-derived colonies compared with colonies derived from untreated protoplasts (Rech et al., 1987). Detailed studies of *Prunus* protoplasts (Ochatt et al., 1988a) showed that electrical enhancement of division was sustained beyond the formation of cell colonies to the callus stage, with electrical pulses of 250 or 500 V eliciting the greatest response, irrespective of pulse duration, up to a pulse length of 50 μsec. In these experiments, tissues derived from both electroporated and control protoplasts were grown over three culture periods, each of 21 days duration, with subculture of 100 mg portions of tissue at each transfer. The pooled data for the fresh weight of tissues derived from electroporated protoplasts were, in some treatments, more than fivefold the fresh weight of controls. In addition, the number of shoots that regenerated from callus derived from electroporated protoplasts was significantly greater than from tissues derived from untreated protoplasts. Shoots produced from electrically treated protoplasts, particularly those exposed to 250 or 500 V, were longer with more leaves than shoots from control tissues. This feature facilitated rooting of such shoots, without the need for an intermediate growth phase to stimulate internode elongation. Interestingly, the rooting response of shoots from electropulsed protoplasts was enhanced, with a maximum response, based on the number of roots per shoot and total root length, at 250 and 500 V. Additionally, the roots of several shoots were branched, unlike those of control shoots. All shoots from electroporated protoplasts rooted readily, whereas 25% of the shoots from control protoplasts failed to root.

Studies of protoplast-derived tissues of the woody medicinal plant *Solanum dulcamara* confirmed that electroporation of isolated protoplasts stimulated organogenesis (Chand et al., 1988). Exposure of protoplasts from cell suspensions to voltages similar to those used for *Prunus* (250–1250 V cm^{-1}) for three successive pulses each of 10–50-μsec duration, again promoted callus growth, biomass production, and shoot formation, such tissues requiring a shorter period in culture before morphogenesis than tissues derived from untreated protoplasts. Shoot formation was well advanced by day 70 of culture in tissues from electroporated protoplasts, whereas tissues from untreated protoplasts had not commenced organogenesis by this time. As in the case of *Prunus*, shoots regenerated from electroporated protoplasts rooted more readily and developed more prolific root systems than shoots from untreated protoplasts. An interesting observation was that shoots initiated on tissues derived from protoplasts electroporated at the highest voltage assessed (1250 V) for 20 μsec were compact and clustered on the surface of the parent tissue. These compact shoots, however, developed into phenotypically normal plants following transfer to culture medium lacking growth regulators.

Whereas shoot regeneration occurred by organogenesis in *Prunus* and *Solanum*, somatic embryogenesis was reported to be the main pathway of plant regeneration in tissues derived from electroporated protoplasts of seedling hypocotyls of *Helianthus annuus* (Barth et al., 1993). Following electrostimulation with two or three pulses of 10 μsec at 1500 V cm^{-1} or 50 μsec at 1200 V cm^{-1}, protoplasts were cultured in agarose-solidified droplets of medium bathed in liquid medium, the latter being replaced every 7 days. Protoplast-derived colonies were transferred to media containing α-naphthaleneacetic acid (2.0 mg l^{-1}), benzylaminopurine (1.0 mg l^{-1}), and gibberellic acid (0.1 mg l^{-1}) to induce morphogenesis. Although only 11 somatic embryos were obtained from 120 protoplast-derived tissues, this result provided important information in relation to the recovery of plants following genetic manipulation of protoplasts of this economically important crop.

In the barley (*Hordeum vulgare*) cultivar Dissa, a beneficial effect of electroporation has been reported on the regeneration of shoots from tissues derived from protoplasts isolated from embryogenic cell suspensions (Mordhorst and Lőrz, 1992). Freshly isolated protoplasts were exposed to both rectangular and exponential electrical pulses. Although increasing the field strength, capacitance and the number of applied pulses beyond a certain threshold decreased protoplast viability and plating efficiency, the regeneration of albino organs (leaves) and albino plantlets from electrotreated protoplasts was stimulated in comparison with controls. The regeneration of albino shoots was not unique to this study in barley, since albinism is commonly associated with plant regeneration from cultured cells of all the major cereals. This phenomenon is, to some extent, dependent upon the plant cultivar and cell line. Consequently, electroporation of protoplasts isolated from other cell suspensions of barley could stimulate the regeneration of phenotypically normal, chlorophyll-containing shoots.

Collectively, these results indicate that electroporation not only stimulates cell division, tissue growth, and morphogenesis, but that electrical pulses also influence plant development in terms of the pathway of regeneration, together with the vigor and phenotype of regenerated shoots.

LONG-TERM EFFECTS OF HIGH-VOLTAGE ELECTRICAL PULSES ON GROWTH OF CULTURED PLANT CELLS

As already discussed, the effects of applying high-voltage, short-duration electrical pulses to freshly isolated protoplasts can be recognized soon after electrical treatment by a stimulation of cell division and, additionally, after several days or weeks at the shoot regeneration and plant development levels. In a long-term investigation of the effects of electroporation on growth and morphogenesis of protoplast-derived cells of *Prunus avium* x *pseudocerasus*, Ochatt et al. (1988c) electroporated (250 V cm^{-1}, three successive pulses each of 30 μsec at 10-sec intervals) before culture protoplasts from cell suspensions. Three months later, tissue derived from electroporated protoplasts was used to initiate a cell suspension (designated ECS), which itself was used as a source of protoplasts for experiments on the long-term effects of electroporation on cell growth and differentiation. Cells of ECS were transferred every 18 days to fresh medium over 16 successive passages. They were harvested for protoplast isolation 14 days after each subculture, when the cells were in exponential growth. The protoplasts from suspension ECS were cultured alongside non-electroporated protoplasts or recently electroporated protoplasts from a control cell suspension. The latter was initiated from root callus at the same time as suspension ECS.

Examination of protoplasts during the first 30 days of culture revealed that the enhancement of growth previously shown to result from electroporation of isolated protoplasts (Ochatt et al. 1988c) was sustained even when protoplasts were isolated from the long-term suspension ECS, itself derived from electroporated protoplasts. Protoplasts from suspension ECS entered division earlier (at 6–9 days) than untreated protoplasts from a control cell suspension (13–17 days), but later than protoplasts from the same control cell suspension that had been electroporated immediately after isolation and before culture (3–5 days). This result was consistent during the 16 passages of suspension ECS. Assessments of plating efficiency also demonstrated that the frequency of division of protoplasts from suspension ECS was higher than for non-electroporated protoplasts from the control cell suspension. Likewise, the number (4–8) of regenerated shoots per protoplast-derived tissue was higher for tissues from protoplasts of suspension ECS compared with the number of shoots (1–4) from tissues of non-electroporated, control protoplasts. The former figure was lower than the number of shoots from protoplasts prepared from control suspensions and electroporated immediately after isolation (9–12).

The results from these experiments demonstrated that electroenhancement of protoplast division and plant regeneration is expressed over a number of subcultures when protoplasts are isolated from an established cell suspension, itself derived from electroporated protoplasts. The growth stimulation exhibited by protoplasts electroporated immediately before culture and by protoplasts isolated from suspension ECS, compared with control, non-electroporated protoplasts, demonstrated that the effects of electrical pulses could not be simply the result of a metabolic change induced immediately after protoplast isolation and resulting, in turn, in an initial increase in endogenous levels of growth regulators. These observations, together with the increase in DNA synthesis detected in *Prunus* protoplasts following electroporation (Rech et al., 1988), may indicate that electroporation influences the expression of genes-controlling differentiation. A sustained capacity to take up components of the culture medium at a higher, more selective, or more efficient rate may have been introduced permanently and transmitted to tissues derived from electroporated protoplasts.

APPLICATION OF ELECTROSTIMULATION OF SHOOT REGENERATION TO THE RECOVERY OF GENETICALLY ENGINEERED PLANTS

As emphasised by Mordhorst and Lőrz (1992), efficient and reliable systems for plant regeneration from protoplasts are essential for somatic hybridization through protoplast fusion and transformation experiments involving direct uptake of DNA into isolated protoplasts. The use of electrical treatment of protoplasts to facilitate the recover of somatic hybrids is exemplified by the work of Ochatt et al. (1988b), who generated hybrid plants between the sexually incompatible top-fruit tree rootstocks, *Pyrus communis* var. *pyraster* (wild pear) and *Prunus avium* x *pseudocerasus* (Colt cherry). Protoplasts of *Pyrus* were isolated from leaves of axenic shoots, whereas those of *Prunus* were prepared from cell suspensions. Before mixing and chemical fusion, separate suspensions of the parental protoplasts were adjusted to a density of 1.0×10^6 mL^{-1} (*Pyrus*) or 4.0×10^6 mL^{-1} (*Prunus*) in their respective culture media and electroporated at 250 V cm^{-1} by discharging a 30-nF capacitor for three successive 87-μsec pulses at 10-sec intervals. The selection strategy for recovering somatic hybrids used a culture medium in which neither of the parental protoplasts was capable of growth in its own right.

The recovery of somatic hybrids was based solely on complementation to growth proficiency, coupled with an anticipated effect of heterosis. Since electroporation had been demonstrated previously to stimulate division of protoplasts of *Prunus* and *Pyrus* (Rech et al., 1987), it was believed that such an electrotreatment before mixing and exposure of the protoplasts to the fusogen [polyethylene glycol (PEG) at high pH] would enhance the production of heterokaryon-derived somatic hybrid cells. Additionally, the stimulatory effect of electroporation on the competence for plant regeneration reported for protoplast-derived tissues of *Prunus avium* x *pseudocerasus* (Ochatt et al., 1988a) was also likely to foster plant regeneration from heterokaryon-derived tissues. It seems likely that electroporation did, indeed, stimulate both the hybrid vigor and shoot regeneration of heterokaryon-derived tissues, enabling unique somatic hybrid plants to be generated between these woody genera.

Electroporation is a routine procedure for inserting DNA into isolated protoplasts for both transient and stable gene expression studies. Several parameters, including the voltage, pulse duration, number of pulses, inter-pulse periods, DNA concentration, and composition of the electroporation solution, influence gene uptake and protoplast viability. In the case of exponentially decaying capacitive discharges, the pulse length is deter-mined by the resistance of the electroporation cell, with solutions of relatively high conductivity facilitating capacitor discharge. The effects of exponentially decaying pulses on protoplast viability have been discussed at length (Saunders et al., 1989; Joersbo and Brundstedt, 1990), with the possibility that capacitive discharges may be more detrimental to proto-plast viability than square-wave pulses. Jones et al. (1993) also provided evidence that electronically defined, square-wave pulses generated by a non-commercial instrument (Jones et al., 1994) stimulated DNA uptake into protoplasts of *Glycine argyrea*, *Medicago sativa*, *M. varia*, and *Stylosan-thes macrocephala*, while conserving protoplast viability. Since the duration of the pulse(s) from the electronic unit was not dependent on the resis-tance of the electroporation solution as the instrument was designed with a high internal resistance, it was possible to use an electroporation solution of low ionic content (low conductivity). This solution contained mannitol for osmoregulation to prevent bursting of the isolated protoplasts and 0.5-mM calcium chloride for membrane stability. Such low ionic solutions also minimize heating and ion uptake effects during permeabi-lization of the plasma membrane. In terms of the recovery of transgenic protoplast-derived plants, the optimum electroporation conditions are those that maximize DNA uptake without compromising such beneficial effects as a stimulation of mitosis of protoplast-derived cells and shoot regenera-tion from protoplast-derived tissues.

An interesting observation of Chand et al. (1989) was that electropora-tion stimulated gene transfer from *Agrobacterium* to freshly isolated protoplasts. Until that time, controversy existed as to whether or not a newly synthesized cell wall was essential for *Agrobacterium*-mediated trans-formation of protoplasts, with a bias toward the view that a new primary cell wall was essential to provide the attachment sites necessary for gene transfer from the prokaryotic to the eukaryotic cells. It was also considered essential that the protoplast-derived cells were mitotically active to enable them to be competent for transformation. Chand et al. (1989) demon-strated that although protoplasts freshly isolated from cell suspensions of *Solanum dulcamara* could not be transformed by cocultivation with *A. rhizogenes*, exposure of protoplasts to two electrical pulses, each of 600 V

cm^{-1} for 2 msec and 15 sec apart, resulted in protoplast transformation frequencies in excess of those normally obtained by cocultivating 3-day old protoplast-derived cells with the same strain of *Agrobacterium*. Freshly isolated protoplasts were mixed with the bacteria 2 hr before electroporation and further incubated with the bacteria for the same period of time post-electroporation. The precise mechanism of this stimulation of gene transfer is not known. Since electroporation stimulates DNA synthesis (Rech et al., 1988), however, such protoplasts may have been more competent for transformation. Electrical pulses may also enhance replication of the T-DNA of the *Agrobacterium* Ri plasmid and subsequent transfer of T-strands from *Agrobacterium* to recipient plant protoplasts. Chand et al. (1989) selected protoplast-derived cell colonies by their antibiotic (kanamycin sulphate) resistance following expression of the neomycin phosphotransferase (*npt*II) gene. Transgenic shoots were regenerated directly from selected colonies and from transformed roots that developed on these antibiotic-resistant tissues.

SUMMARY

Evidence exists from several higher plant protoplast systems to substantiate the hypothesis that short-duration, high-voltage electrical pulses stimulate mitotic division and plant regeneration. Electroporation treatments are simple to initiate and are rapid, unlike long-term low-voltage treatments that necessitate equipment to be run continuously for the duration of any particular treatment. In terms of plant genetic engineering, high-voltage effects have application by increasing the through-put of novel plants, as demonstrated by the somatic hybridization studies of Ochatt et al. (1988b) and the transformation experiments of Chand et al. (1989). In terms of the extent of research on plant genetic engineering using protoplasts as source material, however, most workers appear to ignore the potentially beneficial effects of electrical treatments prior to culture. Clearly, there is a requirement to extend the use of these simple techniques to plant protoplast systems other than those already described in the literature since high-voltage treatments could dramatically stimulate mitotic division and organogenesis in hitherto recalcitrant genera and species. There is also a requirement to evaluate, more extensively, the influence of pulse shape on protoplast division and shoot regeneration.

Recently, other physical and chemical parameters have been shown to stimulate protoplast division, including the use of oxygen-enriched atmospheres (d'Utra Vaz et al., 1992) and oxygenated perfluorocarbon liquids such as perfluorodecalin. In the latter system, the mean plating efficiency of isolated protoplasts of *Petunia hybrida* was increased by 37% during culture at the perfluorodecalin–culture medium interface (Anthony et al., 1994a, 1994b). Supplementation of the culture medium with the non-ionic co-polymer surfactant, Pluronic F-68 at 0.001% (w/v), further increased the plating efficiency of *Petunia* protoplasts to 52% above control. Similarly, the addition of Pluronic F-68 at concentrations below 1.0% (w/v) to culture medium stimulated growth and shoot regeneration of cell suspensions of *Arabidopsis thaliana*, shoot production from cotyledons with attached petioles of *Corchorus capsularis*, and the plating efficiency of protoplasts of *Solanum dulcamara*. Pluronic F-68 also increased the growth of transformed roots of both *Corchorus* and *Solanum* (Lowe et al., 1993, 1994). Additionally, related work, also with *C. capsularis*, has shown that the stimulatory effect of different non-ionic surfactants, including Pluronic F-68, Tween-20, and Triton X-100, is related to their physico-chemical properties, most notably their hydrophilic–lipophilic balance (Khatun et al., 1993). Thus, surfactants may exert their stimulatory effect, at least partially, through permeability changes of the plasma membrane (Lowe et al., 1993), with consequent enhancement of nutrient uptake from the culture medium. In this respect, their mode of action may be somewhat similar to that of electrical pulses, although permeabilization of the plasma membrane by electrical pulses may be transitory compared to surfactant-induced changes. It is conceivable that electrical pulses and surfactants could act synergistically on cultured plant cells and tissues. A combination of these parameters warrants further investigation in a range of plant systems.

References

Anthony, P., Davey, M. R., Power, J. B., and Lowe, K. C. (1994b). Image analysis assessments of perfluorocarbon- and surfactant-enhanced protoplast division. Plant Cell, Tiss. Org. Cult. **38**: 39–43.

Anthony, P., Davey, M. R., Power, J. B., Washington, C., and Lowe, K. C. (1994a). Synergistic enhancement of protoplast growth by oxygenated perfluorocarbon and Pluronic F-68. Plant Cell Rep. **13**: 251–255.

Barth, S., Voeste, D., Wingender, R., and Schnabl, H. (1993). Plantlet regeneration from electrostimulated protoplasts of sunflower (*Helianthus annuus* L.). Botanica Acta **106**: 220–222.

Chand, P. K., Ochatt, S. J., Rech, E. L., Power, J. B., and Davey, M. R. (1988). Electroporation stimulates plant regeneration from protoplasts of the woody medicinal species *Solanum dulcamara* L. J. Exp. Bot. **206**: 1267–1274.

Chand, P. K., Rech, E. L., Golds, T. J., Power, J. B., and Davey, M. R. (1989). Electroporation stimulates transformation of freshly isolated cell suspension protoplasts of *Solanum dulcamara* by *Agrobacterium*. Plant Cell Rep. **8**: 86–89.

Davey, M. R., and Kumar, A. (1983). Higher plant protoplasts—retrospect and prospect. Pages 219–299. In Plant Protoplasts. Internatl. Rev. Cytol. Suppl. 16. Giles, K. L., ed. Academic Press, New York, London.

d'Utra Vaz, F. B., Slamet, I. H., Khatun, A., Cocking, E. C., and Power, J. B. (1992). Protoplast culture in high molecular oxygen atmospheres. Plant Cell Rep. 11: 416–418.

Fromm, M. E., Taylor, L. P., and Walbot, V. (1986). Stable transformation of maize after gene transfer by electroporation. Nature 319: 791–793.

Gupta, H. S., Rech, E. L., Cocking, E. C., and Davey, M. R. (1988). Electroporation and heat shock stimulate division of protoplasts of Pennisetum squamulatum. J. Plant Physiol. 133: 457–459.

Hashimoto, H., Morikawa, H., Yamada, Y., and Kimura, A. (1985). A novel method for transformation of intact yeast cells by electroinjection of plasmid DNA. Appl. Microbiol. Biotechnol. 21: 336–339.

Joersbo, M., and Brunstedt, J. (1990). Direct gene transfer to plant protoplasts by electroporation by alternating, rectangular and exponentially decaying pulses. Plant Cell Rep. 8: 701–705.

Jones, B., Antonova-Kosturkova, G., Vieira, M. L. C., Rech, E. L., Power, J. B., and Davey, M. R. (1993). High transient gene expression, with conserved viability, in electroporated protoplasts of Glycine, Medicago and Stylosanthes species. Plant Tissue Cult. 3: 59–65.

Jones, B. J., Lynch, P. T., Handley, G. J., Malaure, R. S., Blackhall, N. W., Hammatt, N., Power, J. B., Cocking, E. C., and Davey, M. R. (1994). Equipment for the large-scale electromanipulation of plant protoplasts. BioTechniques 16: 312–321.

Khatun, A., Davey, M. R., Power, J. B., and Lowe, K. C. (1993). Stimulation of shoot regeneration from jute cotyledons cultured with non-ionic surfactants and relationship to physico-chemical properties. Plant Cell Rep. 13: 49–53.

Libbof, A. R., Williams, T., Strong, D. M., and Wistar, R. (1984). Time-varying magnetic fields: Effect on DNA synthesis. Science 223: 818–819.

Lowe, K. C., Davey, M. R., Laouar, L., Khatun, A., Ribeiro, R. C. S., Power, J. B., and Mulligan B. J. (1994). Surfactant stimulation of growth in cultured plant cells, tissues and organs. Pages 234–244. In Physiology, Growth and Development of Plants in Culture. Lumsden, P. J., Nicholas, J. R., and Davies, W. J., eds. Kluwer Academic Publishers, Dordrecht, The Netherlands.

Lowe, K. C., Davey, M. R., Power, J. B., and Mulligan, B. J. (1993). Surfactant supplements in plant culture systems. Agro-Food Indust. Hi-Tech. Jan / Feb. 1993: 9–13.

Montane, M. H., and Teissié, J. (1992). Electrostimulation of plant protoplast division. Part 1. Experimental results. Bioelectrochem. and Bioenerget. 29: 59–70.

Mordhorst, A. P., and Lőrz, H. (1992). Electrostimulated regeneration of plantlets from protoplasts derived from cell suspensions of barley (Hordeum vulgare). Physiol. Plant. 85: 289–294.

Neumann, E., Schaefer-Rider, M., Wang, Y., and Hofschneider, P. H. (1982). Gene transfer into mouse lyoma cells by electroporation in high electric fields. EMBO J. 1: 841–845.

Ochatt, S. J., Chand, P. K., Rech, E. L., Davey, M. R., and Power, J. B. (1988a). Electroporation-mediated improvement of plant regeneration from Colt cherry (*Prunus avium* x *pseudocerasus*) protoplasts. Plant Sci. **54**: 165–169.

Ochatt, S. J., Patat-Ochatt, E. M., Rech, E. L., Davey, M. R., and Power, J. B. (1988b). Somatic hybridization of sexually incompatible top-fruit tree rootstocks, wild pear (*Pyrus communis* var. *pyraster* L.) and Colt cherry (*Prunus avium* x *pseudocerasus*). Theor. Appl. Genet. **78**: 35–41.

Ochatt, S. J., Rech, E. L., Davey, M. R., and Power, J. B. (1988c). Long-term effect of electroporation on enhancement of growth and plant regeneration of Colt cherry (*Prunus avium* x *pseudocerasus*) protoplasts. Plant Cell Rep. **7**: 393–395.

Rathore, K. S., and Goldsworthy, A. (1985a). Electrical control of growth in plant tissue cultures. Bio/Technol. **3**: 253–254.

Rathore, K. S., and Goldsworthy, A. (1985b). Electrical control of shoot regeneration in plant tissue cultures. Bio/Technol. **3**: 1107–1109.

Rech, E. L., Ochatt, S. J., Chand, P. K., Davey, M. R., Mulligan, B. J., and Power, J. B. (1988). Electroporation increases DNA synthesis in cultured plant protoplasts. Bio/Technol. **6**: 1091–1093.

Rech, E. L., Ochatt, S. J., Chand, P. K., Power, J. B., and Davey, M. R. (1987). Electro-enhancement of division of plant protoplast-derived cells. Protoplasma **141**: 169–176.

Reiss, M., Jastreboff, M. M., Bertino, J. R., and Narayanan, R. (1986). DNA-mediated gene transfer into epidermal cells using electroporation. Biochem. Biophys. Res. Commun. **137**: 244–249.

Rodan, G. A., Bourret, L. A., and Norton, L. A. (1978). DNA synthesis in cartilage cells is stimulated by oscillating electric fields. Science **199**: 690–692.

Saunders, J., Mathews, B. F., and Miller, P. D. (1989). Plant gene transfer using electrofusion and electroporation. Pages 343–354. In Electroporation and electrofusion in cell biology. Neumann, E., Sowers, H. E., and Jordan, C. A., eds. Plenum Press, New York.

Shillito, R. D., Saul, M. W., Paszkowski, J., Müller, M., and Potrykus, I. (1985). High efficiency direct gene transfer to plants. Bio/Technol. **3**: 1099–1103.

Takahashi, K., Kaneko, I., Date, M., and Fukuda, E. (1986). Effect of pulsing electromagnetic fields on DNA synthesis in mammalian cells in culture. Experientia **42**: 185–186.

Toneguzzo, F., and Keating, A. (1986). Stable expression of selectable genes introduced into human hematopoietic stem cells by electric-field mediated DNA transfer. Proc. Natl. Acad. Sci. USA **83**: 3496–3499.

Index